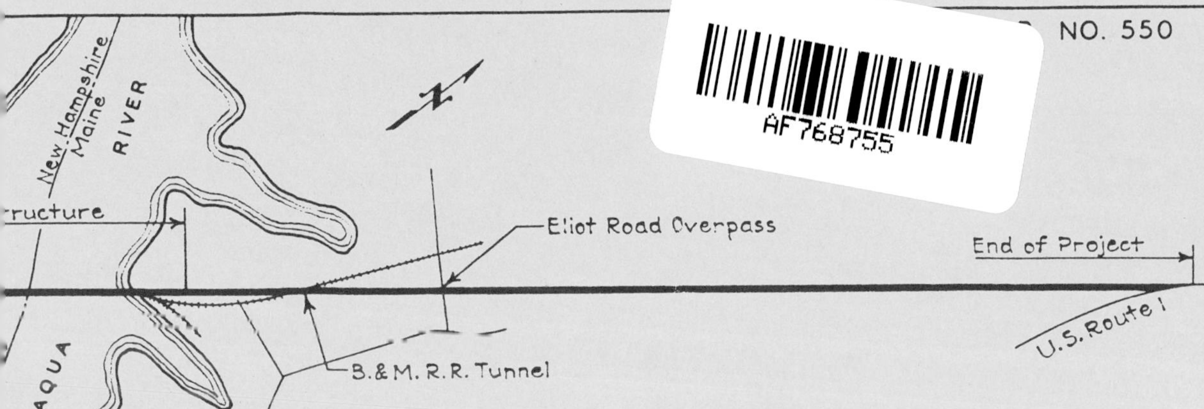

ROJECT

3000

SPECIFICATIONS

For Railway and Combined Structure : A.R.E.A.-1935.
For Highway only : A.A.S.H.O.-1935.
For Machinery: A.R.E.A.-1922 as amended by Harrington & Cortelyou.

LIVE LOADS

For Railway : E72
For Highway: H 20

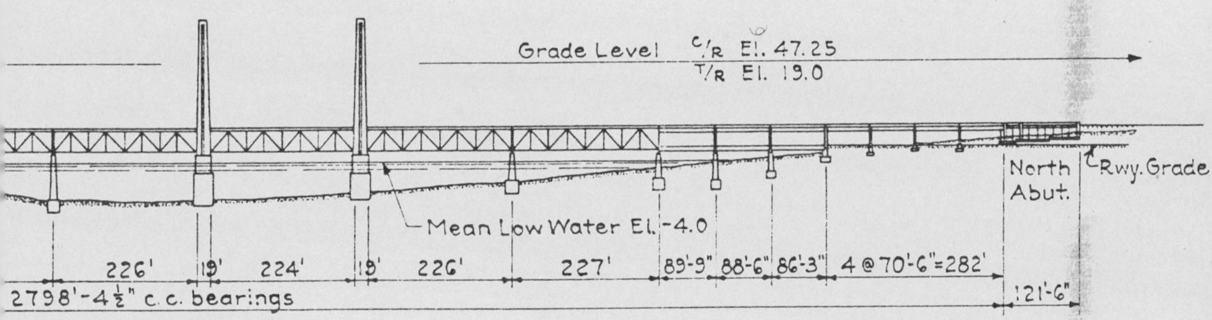

ATION OF MAIN STRUCTURE

0 100 200 250
Scale in feet

GENERAL DATA

Construction Cost of Main Structure $1,872,300
Construction Cost of Approach Work $1,095,000
Total Cost of Project, including Engineering, Right-of-Way, etc. approx. $3,220,000.

Total Length of Project approx. 4.3 miles.

Constructed 1940.

PROACH WORK

	20	acres
	187,850	cu.yds.
	88,000	cu.yds.
	80,400	cu.yds.
	100,420	sq.yds.
es:		
	427,000	lbs.
	489,000	lbs.
otings	1,946	cu.yds.
	4,777	cu.yds.
	832,000	lbs.
	Lump Sum	

MAINE-NEW HAMPSHIRE INTERSTATE BRIDGE
OVER THE PISCATAQUA RIVER
PORTSMOUTH, N.H. TO KITTERY, ME.

HARRINGTON AND CORTELYOU
CONSULTING ENGINEERS
KANSAS CITY, MISSOURI.

SEPT. 1949 DATA SHEET

to: Kathleen Kristiansen 1/30/07
from: Jay

The Sarah Mildred Long Bridge

A History of the Maine-New Hampshire Interstate Bridge from Portsmouth, New Hampshire, to Kittery, Maine

The Sarah Mildred Long Bridge

A History of the Maine-New Hampshire Interstate Bridge from Portsmouth, New Hampshire, to Kittery, Maine

Woodard D. Openo

The Portsmouth Marine Society
Publication Thirteen

Published for the Society
in cooperation with
The Maine-New Hampshire Interstate Bridge Authority by

Peter E. Randall
PUBLISHER

Produced by
Peter E. Randall Publisher
Box 4726, Portsmouth, NH 03801

The Portsmouth Marine Society
Box 147, Portsmouth, NH 03801

Frontispiece—Interstate Bridge and Kittery, Maine, end of bridge. Bridge operator's house as originally constructed with one story. C.1948. Photograph by Douglas Armsden

Front end sheet—Map: U.S. Route 1 Bypass and Interstate Bridge; Portsmouth, N.H., to Kittery, Me., as originally built. Harrington and Cortelyou, Consulting Engineers, Data Sheet 1 of 9, September 1949.

Rear end sheet—Map: U.S. Route 1 Bypass-Portsmouth, N.H., to Kittery, Me., with 1950 Portsmouth Traffic Circle and connected 1956 Spaulding Turnpike; 1950 U.S. Interstate 95 Highway, with its 1972 revisions and 1972 Piscataqua River (High Level) Bridge, west of the Interstate (Maine-N.H.) Bridge. U.S.G.S. Topographic Map, Portsmouth Quadrangle, N.H.-Maine, 1956, photo revised 1981.

Library of Congress Cataloging-in-Publication Data

Openo, Woodard D., 1943-
The Sarah Mildred Long Bridge.

(Publication / Portsmouth Marine Society; 13)
Bibliography: p.
Includes index
1. Sarah Mildred Long Bridge (Kittery, Me. and Portsmouth, N.H.) 2. Maine-New Hampshire Interstate Bridge Authority--History. I. Title. II. Series: Publication (Portsmouth Marine Society); 13.
TG25.K57064 1988 388.1'32'097426 88-25592
ISBN 0-915819-12-0

Other Portsmouth Marine Society Publications:

1. *John Haley Bellamy, Carver of Eagles*
2. *The Prescott Story*
3. *The Piscataqua Gundalow,*
   Workhorse for a Tidal Basin Empire
4. *The Checkered Career of Tobias Lear*
5. *Clippers of the Port of Portsmouth*
   and the Men who built Them
6. *Portsmouth-Built*
   Submarines of the Portsmouth Naval Shipyard
7. *Atlantic Heights*
   A World War I Shipbuilders' Community
8. *There are No Victors Here*
   A Local Perspective on the Treaty of Portsmouth
9. *The Diary of the Portsmouth, Kittery*
   and York Electric Railroad
10. *Port of Portsmouth Ships and the Cotton Trade*
11. *Port of Dover*
    Two Centuries of Shipping on the Cochecho
12. *Wealth and Honour*
    Portsmouth During the Golden Age of
    Privateering, 1775-1815

Foreword

THE MAINE-NEW HAMPSHIRE INTERSTATE BRIDGE AUTHORITY is pleased to make the following publication available as a part of its 50th Anniversary celebration. The Authority was created by Acts of the Maine and New Hampshire Legislatures in 1936 and 1937. It is made up of two appointed members each from Maine and New Hampshire with the Commissioners of each State's Department of Transportation serving in an ex-officio capacity and alternately assuming the chairmanship. The compact between the States was approved by the Congress of the United States on June 16, 1938. This bridge, 2798 feet in length with a 224 foot center lift span, was opened to vehicular traffic on November 8, 1940. It was designed by Harrington and Cortelyou, Inc., of Kansas City, Missouri.

Last year, the bridge was formally named the Sarah Mildred Long Bridge in honor of the singular and extraordinary 50 years of exclusive dedicated service given by our current Executive Secretary, dating back to the inception of the Authority.

The bridge has long stood as a symbol of the warm partnership and friendly cooperation between the two neighboring New England states.

This book now stands as a lasting tribute to all the men and women who have shared in the work of the Authority over these many years.

Dana F. Connors, Chairman
Roger W. Staples
Rita A. Perry
Wallace E. Stickney
Robert A. Allard
Harold J. Silverman

To my parents, who gave me an appreciation
for people and history,
and to Sarah Mildred Long.

Contents

Acknowledgments

I WISH TO EXPRESS my appreciation for the help given me by many people in writing this history of the Sarah Mildred Long (Interstate) Bridge for the Maine-New Hampshire Interstate Bridge Authority. Without exception, present and past members of the Authority and its employees, members of the transportation departments of the two states, people connected with navigation in Portsmouth Harbor and those with local knowledge of Kittery, Maine, and Portsmouth, New Hampshire, have been generous with their time and efforts. Special thanks are due to John Moore of the Bridge Design Office of the New Hampshire Department of Transportation (N.H. D. O. T.); Richard V. Giles, Bridge Construction Superintendent, N. H. D. O. T., Ernest Searles, Maintenance Engineer, N. H. D.O.T.; David H. Stevens; Everett B. Barnard of the Bridge Maintenance Section of the Maine Department of Transportation; Willis G. Grinstead of the firm of Harrington and Cortelyou, Consulting Engineers; Edwin M. Newman of Corpus Christi, Texas, engineer; H.D. Peoples of San Marino, California, engineer; Captains Richard Holt and Norman Y. Holt; Douglas Armsden, photographer; Chandler B. Cobb, railroad enthusiast; Hiram A. Young, Supervisor of Maintenance for the Bridge; Mrs. Carol Drew of the Bridge Maintenance Office; Leland Riley, Jr.; Ms. Catherine P. Jarvis; the late Norman Clark; Ray Brighton; and, of course, Miss Mildred Long, mainstay of the Bridge Authority. Present and past bridge operators and gatemen (James Abel, Robert Williams, the late James Howard, Charles Starkey and Clarence Foley in particular) have shared their vast knowledge of the bridge and the locality, and most are mentioned somewhere in the work. Space does not allow acknowledgment of all who helped, and some who provided key information or led to outstanding sources. They have my appreciation.

Finally, while I doubt that I have done credit to the work and knowledge of these people, I hope that this brief history will commemorate something which is worthy of recognition and provide a starting point for future historians.

Woodard D. Openo
Somersworth, N.H.
July 30, 1988

Badger's Island; Kittery, Maine, with Portsmouth-Kittery ferry at dock. Old railroad bridge in background. C.1923 from Memorial Bridge tower. Photographer facing west. Postcard courtesy of Leland Riley, Jr.

Old Railroad Bridge. C.1934. Photographer facing south toward Portsmouth. Kittery Junction depot at center right, at end of bridge. Photograph courtesy of Leland Riley, Jr.

Introduction

MOST PEOPLE DRIVE ACROSS BRIDGES without thinking much about them. They might look at the view or, more likely, remain intent on their destination. Bridges, however, have great symbolic value, a fact recognized by the ancient Romans and by others long before them. They represent beginnings and endings; they connect the past and the future. Geographically, they are found at strategic points of communication. Technically, they have frequently called forth the best civil engineering of the time. The Sarah Mildred Long Bridge is all of these things; it is also a story of continuous public service by many individuals over a half-century. It is a great story, one which deserves to be better known.

For over three and a half centuries of recorded history, the Piscataqua River, a tidal estuary, has served as a means of communication between the Atlantic Ocean and the interior of the Maine-New Hampshire Seacoast area. For an equally long period of time, settlers have sought to overcome the barrier to land transportation posed by the river. Newspaper accounts which appeared during planning and construction of the Interstate Bridge elucidate its function as an improved facility within the historical context of local bridges and their connected roads; clearly, the general public's understanding of this context was assumed by the reporters, who compared the new bridge with the ferry crossings and earlier bridges. Everyone knew how important the river was in the life of the Seacoast.

One of the earliest efforts to span the river was the building of a bridge from Fox Point in Newington, New Hampshire, across to Durham in 1794. Then, in 1822, the bridge from Portsmouth to Kittery was opened. Much later, in 1873, a railroad bridge was built

Miss Sarah Mildred Long, Executive Secretary of the Maine-New Hampshire Interstate Bridge Authority, with plaque rededicating the bridge in her honor. Photograph taken June 5, 1987, in the Bridge Office. Above the plaque is an architectural rendering of the bridge, executed when it was designed, by T.J. Cambern, Chief Draftsman with Harrington and Cortelyou, Consulting Engineers.

from Newington to Hilton's Point in Dover. The Portsmouth-Kittery bridge was modified to allow rail traffic when the railroad was put through in 1842, at which time the Kittery Junction station was built at the north end. Trains crossed this bridge until the construction of the Interstate (Long) Bridge in 1938-40. The old Portsmouth-Kittery bridge was joined in 1923 by the Memorial Bridge between the two municipalities, after which vehicular traffic lasted only another year or so. The railroad history of the bridge ended on September 10, 1939, when a locomotive and tender plunged through one of its spans. The bridge continued to serve an important purpose, though, as it provided the main construction platform for the new bridge, which was dedicated on November 8, 1940. Temporarily rerouted through Dover, New Hampshire, the railroad traffic across the new bridge diminished after World War II. Originally called the Portland, Saco and Portsmouth Railroad, the line became part of the Eastern Railroad, and, finally, part of the Boston and Maine Railroad, bought out in 1973 by Guilford Transportation. As the Eastern Division of the Boston and Maine, the line operated until September 27, 1952. Now the only trains which cross the bridge carry freight to the Portsmouth Naval Shipyard in Kittery. Other developments that affected the Interstate Bridge were the construction of I-95 as part of the Interstate Highway System in 1950 and the Piscataqua River Bridge (the "High Level Bridge") in 1972, the latter being located a short distance upstream from the Long Bridge.

These are the bare facts of the history of the bridge, but its life is better chronicled by the individuals who have worked on it or travelled by it—the past and present Interstate Bridge Authority members and employees, and local residents. One of these individuals was Miss Sarah Mildred Long, employed by the Authority as Executive Secretary and in other capacities for fifty years, beginning in 1937. She continues to work for the Authority as a consultant. This remarkable woman represents the connection of the region's present with its past, the bridge being renamed in her honor in the anniversary year of 1987. The Bridge Authority felt that due to Miss Long's having worked for the Authority for fifty years, since its inception and virtually her entire working life, the honor was well-deserved. Voted unanimously by the Authority, it was approved by the legislatures of Maine and New Hampshire.

One of the best ways of experiencing the history and lore of the bridge is to stand in the operators' house on the Interstate Bridge and talk to one of the older operators. He will undoubtedly tell you about the pulpwood from York that crossed the old bridge, the view of Nobles Island, the schooner barges, the Atlantic Shipyard and all the rest. Of course, this lore is a combination of truth and fiction—some of these things happened on other bridges and some never happened at all (It's been told, for example, that a man fell into concrete that was being poured in a Memorial Bridge pier, and because his fellow workers couldn't get him out, they just kept pouring).

In any case, although the Interstate Bridge quickly became part of the region's mythology, the facts of the bridge's construction and operation are interesting enough; it is a structure which has touched many lives and inspired much devotion over the years. It has also witnessed, and been affected by, a half-century of American history. As Dana Conners, present Chairman of the Authority, said, "Over the years, the bridge has become a person."

In terms of engineering, architecture and financing, the Long Bridge offers many points of interest. Technically, this vertical lift bridge incorporated elements that were important innovations. Its construction was made problematic by swift tidal currents, frigid winter temperatures and excavation of piers as deep as 90'. Architecturally, the new bridge was noteworthy, and its importance is more easily recognized now, with the passage of time, than it would have been when built. Financially, the history of the bridge illustrates how a project of this type was paid for in an era of great demands on government resources.

Old Railroad Bridge; Portsmouth, N.H., to Kittery, Maine. Opened in 1822, railroad use began in 1842. Photographer facing south toward Portsmouth. Taken before 1924 when the vehicle part was still in use. Ray Tobey Collection.

I *Before the Bridge Was Built*

THE PORTSMOUTH-KITTERY BRIDGE which preceded the Long Bridge carried wagons and foot traffic until 1842, when the Portland, Saco and Portsmouth Railroad crossed the bridge. Vehicular traffic continued along with the railroad until the Memorial Bridge (commemorating soldiers and sailors from the two states who served in World War I) was opened in 1923.

The vehicle and pedestrian side of the old pile bridge continued for another year. The toll collectors on the old bridge were reputed to have grown rich, despite charging pedestrians a toll of two cents, while the *Portsmouth Herald* complained about the place's poor lighting and evil reputation. Although the bridge contained literally thousands of pilings, inserted in a vain attempt to achieve stability, its deck was reported to rise and fall like the sea, and sway back and forth with the tides, providing a seemingly dangerous and occasionally terrifying experience for late-night travellers.

This 1822 pile bridge was located sixty feet upstream from the present Interstate (Long) Bridge. The old bridge was 1600' long, with an arch near its southern (Portsmouth) end, and a draw span of fifty feet toward the other end. The railroad maintained a pen in the river where replacement pilings were kept, ready for use. This was located in a shallow area near the south shore and just west of the present bridge; the location can be clearly seen in an aerial view. A number of the pilings from the bridge remain visible as well as the stone abutment at the Kittery end of the bridge. The planked coverings of the pony trusses (i.e. trusses without top chords connecting the sides above the roadway) served to protect the structure from the weather, as did the enclosure of covered bridges.[1]

The old bridge experienced several notable railroad accidents in its history, one of them on July 18, 1904, when an engine, tender and five cars of a northbound freight train went through the draw, which was open to allow passage upriver of the schooner Sadie A. Kimball to obtain a load of bricks; the railroad cars floated away and two were found on Rye Beach, while two steers, sheep and pigs came ashore at Badger's Island in Kittery and (so it is said) were given homes by local inhabitants. A *Portsmouth Herald* article of the following day mentions that there was a similar accident off the bridge about twenty-five years previously. By the time the Interstate Bridge was being built in the late 1930s, the old bridge was considered so unsafe by the railroad that someone was assigned to walk it every half hour. Although it was inspected annually by a diver, it is reported that one of the divers was accustomed to imbibing (vodka) liberally before descending, and frequently fell asleep when he reached his destination. While the Interstate Bridge was being built, on September 10, 1939, a span on the old bridge collapsed under a southbound passenger train, dropping the engine and tender into the river and killing the engineer and fireman, respectively John Beattie of Somerville, Massachusetts, and Charles D. Towle of Exeter, New Hampshire; the tide was incoming and Towle's body was found by Capt. Shirley Holt, Jr., Harbor Master, who headed the rescue party. Although at the time repairs were discussed, they were never carried out and trains were diverted to the Western Branch of the Boston and Maine Railroad (through Dover, N.H., and South Berwick, Maine). The old bridge was used as a construction platform for the Interstate Bridge and appears in construction photographs. It was demolished soon after the Interstate Bridge was opened on November 9, 1940, but came back to haunt builders in the 1960s, when the new bridge was altered to include a movable small craft span. They ran into literally hundreds of pilings in the path of the channel for the new span. In this regard, it should be noted that the "lock splice" piles used in the bridge were believed by Frank A. MacIntyre of Malden, Massachusetts (the diver who made the final inspection of the bridge in April 1939; he was not the diver abovementioned) to be the only ones ever used in construction work. By this, it can be surmised that the piles had to be spliced due to the considerable depth required.[2]

Indeed, the prospect of relief from this Dickensian creation was greeted with enthusiasm by the two communities, who justifiably regarded the new bridge as a step into the future. The volume of automobile traffic had increased rapidly after World War I, overwhelming the Memorial Bridge on summer weekends only a few years after it

Piscataqua River traffic; a four-masted schooner with Nobles Island, Portsmouth, N.H., in the background. She was going to Newington to pick up fertilizer from a plant which produced it there; the dock was where the Schiller Power Plant is today. Photographer facing southwest from Memorial Bridge. 1920s. Photograph and information courtesy of Captain Norman Y. Holt.

Old Railroad Bridge. 1939. Photographer facing west. Photograph by H.D. Peoples. As an engineer coming from the Midwest, Mr. Peoples was naturally interested in this early bridge.

Old Railroad Bridge. 1939. Photographer facing southwest. Photograph by H.D. Peoples.

Old Railroad Bridge, Kittery end. Photographer facing northwest. Photograph by H.D. Peoples.

Old Railroad Bridge. C.1939. Photographer facing west. Trusses are covered to protect them from the weather, as in a covered bridge. In their later years, they also sheltered gentlemen who enjoyed their wine, and dropped the empty bottles on passing boats; this was highly irritating to local fishermen, at least one of whom threatened to shoot the whole bunch! History related by E. Searles, photograph by H.D. Peoples.

Interstate Bridge. South approach, with Nobles Island, Portsmouth, to left, and North Millpond in background. The shadow to the right of the bridge is in the approximate location of the old railroad bridge and the shallow area to the right of its land connection is where the pen for pilings was located. September 4, 1987. Photographer facing south. Photograph by author.

Old railroad bridge. 1939. Photographer facing south from draw. Photograph by H.D. Peoples.

Engine 3666 and tender no. 2024 of the Boston and Maine Railroad; as part of train 246, they fell through the old bridge on September 10, 1939, southbound from Portland. Shown here in Portsmouth. They remain in the river, but were moved out of the channel. Divers have stripped them of souvenirs. Ray Tobey Collection.

was built (the fact that the State of Maine stationed cornborer inspectors on the bridge for several years did not help matters). In addition, it was felt that a better, more economical bridge, which did not require continuous reconstruction, would help keep the Boston and Maine Railroad Eastern Division in Portsmouth. Also recognized was the size limitation placed on ships passing upriver by the narrow draw of the old pile bridge.

Obviously, the new bridge brought many changes to the landscape, but the Route 1 Bypass in Portsmouth and Kittery, built as part of the project, brought even more, because it led to the division of old farms, the cutting-off of streets and the moving or demolition of houses, barns and other buildings. At the Kittery end of the bridge, the bypass forever altered the landscape around Kittery Junction, the railroad depot located just north of the end of the old bridge (and directly in the path of the new road). This area was populated by the Holts, Williamses, Stapleses, Rileys and others, most of them connected with the railroad (Walter M. Staples was foreman of the B. & M.

Old Railroad Bridge and Caisson no. 17 with Snare Corp. barge after the September 10, 1939 accident. The caisson had floated up against the bridge, and the question was whether this caused, or was the result, of the accident. Note the ebbing tide. September 13, 1939. Photographer facing north. Photograph by H.D. Peoples.

Kittery Junction depot of the Boston and Maine Railroad. C.1920. The buggy has just left the bridge; the house behind it is on the corner of Bridge Street and remains today. Photograph courtesy of Chandler B. Cobb.

Holt house; Kittery Junction; Kittery, Maine. C.1920s. Photographer facing west. Norman Y. Holt's father, Captain Shirley Holt, and aunt seated on the porch. This house was demolished in 1952 for the Oak Terrace loop, constructed as part of bypass improvements. Photograph courtesy of Captain Norman Y. Holt.

Railroad construction crew on the old bridge and the Dover Point Bridge), the tugboats or the bridges. The Holts were tugboat operators far back into the nineteenth century and the Williamses also operated tugboats and were bridge operators. Even before the 1822 bridge was built, Kittery Junction was the site of a ferry landing. According to Robert Williams, retired bridge operator, the stone pier for that ferry can still be seen at low tide, on the Kittery shore just upstream from the stone abutment for the old bridge. That abutment, incidentally, was very close to the Kittery Junction depot, as revealed by photographs of the building, so foundations of the depot may very well exist between the present bridge and Oak Terrace. In addition, a small Greek Revival house visible in old photographs is said to have been connected with the ferry.[3] The little community centered on the depot can today be reconstructed only with the help of maps and photographs; until 1939, it was an important feature of the Kittery landscape, with roots going back into the eighteenth century.

Kittery Junction depot. C.1935. Photographer facing north. The Hurd house, behind to the left, is said to have preceded the 1822 bridge, having been connected with the ferry. Along with the Holt house, it survived the Interstate Bridge construction but succumbed to the Oak Terrace project in 1952. Photograph courtesy of Leland Riley, Jr.

Kittery Junction; Kittery, Maine, with Old Railroad Bridge in foreground. Depot in center. 1939. Photographer facing north. Photograph by H.D. Peoples.

Mark Dennett School; Kittery, Maine. C.1920. This was located northwest of the intersection of the Route 1 Bypass and Eliot Road (Rte. 103). It later served as the American Legion Hall, and was moved to Old Post Road, south of the athletic field (where it remains) when the bypass was built. Photograph courtesy of Leland Riley, Jr.

Interstate Bridge construction. C.1939. Photographer facing south. Photograph courtesy of Leland Riley, Jr.

Interstate Bridge construction. C.1940. Photographer facing south. Photograph courtesy of Leland Riley, Jr.

North of Kittery Junction were Butler's Crossing and the old schoolhouse, flanking Route 103; the schoolhouse would be moved to Old Post Road. Farther north were farms, including some pasture land as well as many blueberry fields and thickets. The Alfred Sterling and Samuel Seward farms were in the area where the Kittery Traffic Circle would be built in 1942.

The existing route of the Eastern Division of the Boston and Maine Railroad largely determined the location of the new bridge. Portsmouth was anxious to keep the railroad service and obviously wished to avoid placing unnecessary expenses, such as moving the line, on the B. & M. Thus, one of the arguments made for the Interstate Bridge was that maintenance costs would be lower. In all respects, keeping the land connections of the new bridge in roughly the same locations would make construction simpler and cheaper. The location of the connecting Route 1 Bypass was largely dictated by these considerations, as well by the fact that the bridge site offered the most direct connections to the existing U.S. Route 1 in Maine and New Hampshire. The Portsmouth section of the bypass would curve around the area known as "the Creek," thereby largely avoiding the built-up parts of the city. The intention of the bypass was to provide an alternative to crowded Portsmouth streets for

U.S. Route 1 Bypass in Portsmouth, N.H., with Interstate Bridge to the right. C.1948. Photograph by Douglas Armsden. Photographer facing west.

through traffic; travellers would have to stop only for the traffic light at Woodbury Avenue in Portsmouth and to pay the toll at the bridge (unless the bridge was up!).

The route of the railroad between Kittery Junction and South Berwick, Maine, can be traced northwestward from the Interstate Bridge or, more accurately, from the surviving stone abutment of the 1822 pile bridge. From there it made a loop eastward (overshadowed by the north end of the later Interstate Bridge) and then westward, the latter part following the line of the railroad tunnel which would be built under the Route 1 Bypass. Incidentally, the tunnel, now unused, still exists. The railroad passed through Butler's Crossing, on the south side of Route 103, starting a long diagonal (in relation to the later U.S. Route 1 Bypass and Maine Route 236), joining the subsequent path of Route 236 at Fernald's Crossing, the east end of Fernald Road in Kittery. From there, it headed toward South Berwick, Maine, until it reached Maine Route 91, where it diverged from the later Route 236 and headed northward.[4] The accompanying aerial pho-

Kittery, Maine, Traffic Circle, U.S. Route 1; Route 1 Bypass. Photographer facing southwest. January 1955. Photograph by Douglas Armsden.

Interstate Bridge with land connections. Photographer facing southwest. January 1955. Photograph by Douglas Armsden.

U.S. Route 1 Bypass in Portsmouth, N.H., with Interstate Bridge and Kittery, Maine, in background. Harvey's (Coakley) Brook goes from left to right across lower part of picture, intersecting with Boyd Road, which runs northeastward. Visible are the Woodbury Avenue grade-level crossing and overpasses at New Franklin School and Maplewood Avenue, as one heads toward the bridge. To the right of the Boyd Road / Woodbury Avenue intersection is the Levi Woodbury Mansion, demolished c.1950s for housing for the elderly. Photographer facing north. C.1948. Photograph by Douglas Armsden.

North end of Interstate Bridge with U.S. Route 1 Bypass heading north to connect with I-95 Highway. Visible is the railroad tunnel under the bypass, with the abandoned railroad grade pointing toward the upper left corner of the picture. Kittery Traffic Circle in upper right center, with Route 1 running from it to middle right edge of picture. January 1955. Photographer facing north. Photograph by Douglas Armsden.

tographs, taken in the 1950s, clearly show the railroad near the bridge, as do topographical maps of that era. Construction of the Route 1 Bypass between 1938 and 1940 not only altered the landscape physically by introducing an elevated barrier between some landmarks and causing others to be moved or demolished, it also changed the perception of the landscape, which had centered on the railroad for a century, to one dominated by the new highway. Also relegated to obscurity were the old farms and homesteads along the Bypass.

On the Portsmouth side, the effect was much the same. The Route 1 Bypass crossed some of the earliest homesteads in New Hampshire in the section known as Christian Shore, and a number of farms to the southward. Here, too, the new road would cut off streets, cause buildings to be moved and demolished, and generally alter the

Interstate Bridge with Nobles Island in lower left corner of picture. National Gypsum Co. Plant in center left. Photographer facing west. January 1955. Photograph by Douglas Armsden.

perception of the landscape as well as its physical character. Affected would be the Joseph Cohen and Cornelius Coakley farms as well as Coakley Brook (also known as Hodgdon's or Harvey's Brook), which was straightened into a 12' channel to pass under the road. The Coakley Farm was a dairy farm, as was the Meadow Brook Farm, later the site of the Meadowbrook Motel. The Harvey Brook structure appears in Authority records as a "cattle underpass," so it must also have served that function when first built.[5]

Harvey's (or Coakley) Brook flows from Harvey Pond, a place where ice was once cut, into the head of the North Millpond in Portsmouth. Continuing southeastward, the road would cross land of Mrs. Mary Sullivan, no longer a working farm by 1940, and come to Boston and Maine Railroad land. This land was bounded on the west by the tracks of the Portsmouth and Concord Railroad (operated by the B. & M.) and on the east by the Boston and Maine Eastern Division tracks, which ran into Portsmouth behind the Morley Button Company on Islington Street (The Portsmouth and Concord tracks would be moved to run alongside the others, so that only one railroad overpass would be required; the bridge is said to have been the first designed by Frank M. "Bill" Cortelyou Jr.).

U.S. Route 1 Bypass in Portsmouth, N.H. Intersection of bypass with Woodbury Avenue, which runs diagonally from lower right corner of picture; Howard Johnson's Restaurant (opened on November 9, 1940, the day after the Bridge dedication) is visible to the right of the intersection—its original location. Boyd Road crosses the center of the picture from left to right. This shows the area of the Portsmouth Traffic Circle immediately before its construction. I-95 Highway approaching from upper right corner. C.1948. Photographer facing southeast. Photograph by Douglas Armsden.

A graphic record of the properties and topographical features which preceded the Interstate Bridge and the Route 1 Bypass is preserved in the 1938 right-of-way plans drawn up for the Bridge Authority, which had to condemn and purchase the land needed for the new road, and work out agreements with the railroad concerning its property. For example, between the Portsmouth Traffic Circle and the south end of the Bypass, the new road would obliterate a pond on the Schurman property once used for harvesting ice. It would also obscure the site of the trolley crash connected with the blowing-up of

Henderson's Point on July 22, 1905. The collision of two trollies occurred at the point where the Route 1 Bypass now crosses below Middle Road, in front of the James W. Schurman house at 472 Middle (the house was moved back 25'or 30' when the overpass was constructed c.1939).

Construction of Memorial Bridge between Portsmouth, N.H., and Kittery, Maine (opened August 17, 1923). Floating the south span into place. c.1922. Photographer facing north, toward Badger's Island. Photograph courtesy of Captain Norman Y. Holt.

Memorial Bridge construction. Floating the lift span into place. The three main spans were set on cribbing on barges, towed into place by tugboats, and settled onto the piers when the tide ebbed. Captain Shirley Holt at the top corner of the span, directing his tugboats. C.1922. Photographer facing north. Photograph courtesy of Captain Norman Y. Holt.

II *The Plan*

SOON AFTER THE MEMORIAL BRIDGE opened in August of 1923, monumental traffic jams began tying up downtown Portsmouth, particularly on holiday weekends in the summer.

By the beginning of 1930, residents were clamoring for a solution to the problem. On January 1, the new Mayor of Portsmouth, Fernando W. Hartford, mentioned the idea of the bridge in his inaugural address. The concept was supported by Lawrence Whittemore of the Boston and Maine Railroad and, later, by editorials on August 3 and September 13, 1933, in the *Portsmouth Herald*. As noted, the proposed bridge was regarded as a way of easing traffic congestion in Portsmouth (by taking 'through traffic' on a highway around the city), keeping the Boston and Maine Eastern Division in Portsmouth and, in general, promoting business.[7]

The idea of a new bridge received more serious consideration at a meeting held in Mayor Robert Marvin's office at the Portsmouth, New Hampshire, City Hall on July 20, 1935. The meeting was called by Harold J. Lockwood, U.S. Works Projects Administration director for Maine, New Hampshire and Vermont, to consider building a second railroad and highway bridge over the Piscataqua River. According to a *Portsmouth Herald* article of the same day, the bridge would relieve Portsmouth of congestion caused by heavy summer traffic and would insure the continuation of the Eastern Division of the Boston and Maine Railroad. Besides Lockwood, those present included Boston and Maine engineers (who presented drawings of a bridge in two locations); Lawrence Whittemore and Jeremy Waldron representing the Railroad; Frederic E. Everett of the New Hampshire Highway Department and his assistant, Mr. Kimball; Chief Engineer Burroughs of Maine and members of his staff; the chief engineer of the Boston and Maine, and Maine Central Railroad and staff members; Former Mayor F.W. Hartford of Portsmouth, Vice President of the Portsmouth Civic Council and Chamber of Commerce; Harold Eager, Secretary for Governor Bridges of New Hampshire; Rockingham County, N.H., Commissioner Simes Frink and others. At the meeting, the possibility of making an application to the federal government for the construction of a combined railroad and highway bridge between Portsmouth, New Hampshire, and Kittery, Maine, was discussed.[8] An editorial two days later in the *Portsmouth Herald* warmly supported the project.

Subsequently, legislation was introduced in the Maine and New Hampshire state legislatures to provide for the creation of an Interstate Bridge Authority; a bill was passed by the Seventy-fifth U.S. Congress as S.2661, granting "the consent of Congress to a compact entered into by the States of Maine and New Hampshire for the creation of The Maine-New Hampshire Interstate Bridge Authority" and authorizing the Authority "to construct, maintain, and operate a toll bridge across the Piscataqua River at or near Portsmouth, State of New Hampshire."

Maine Legislative Document 18, Section 1, of the Private and Special Laws of 1937, establishes the compact between the two states, giving the reasons for building a bridge. The compact begins by explaining the reasons for the new bridge, citing the excessive congestion in Portsmouth and Kittery as evidence that the Memorial Bridge was not adequate for the vehicular traffic between the two points. The document goes on to say that the wooden pile bridge which served as the sole facility for railroad traffic over the river was antiquated and obsolete, and that its draw was not large enough to permit passage of steamers and vessels of broad beam—a defect which limited the development of industry along the river banks. The solution to these problems, of course, was a new bridge with improved rail and vehicular facilities and a wider draw.

This bridge would "result in great economy and benefit not only to the states of Maine and New Hampshire but to the nation and [would] require the cordial cooperation of the states of New Hampshire and Maine...." Therefore, in the compact, the two states pledged to cooperate in the planning, execution and construction of the bridge; they defined the Maine-New Hampshire Interstate Bridge Authority as consisting of 6 members—3 residents of the State of New Hampshire and 3 residents of the State of Maine.

The New Hampshire legislation setting up the Authority was comprised of two acts. Chapter 4 of the Special Session of 1936 of the New Hampshire Legislature set up the Authority, while Chapter 86 of the 1937 Session prescribed the manner of appointment, terms of office and compensation of the members. The New Hampshire legislation included an important addition to the Maine legislation: Section 14 of Chapter 4 of the 1936 Session forbade the construction of any competing bridge in the vicinity until the bonds were paid off. Chapter 86, Section 1 of the 1937 New Hampshire Legislature, stated that no more than two members of the Authority from New Hampshire should belong to the same political party, and Section 2 that members other than the Highway Commissioner should be appointed initially for terms of two and four years respectively, and thereafter for terms of four years.

On June 3, 1937, the first meeting of the Maine-New Hampshire Interstate Bridge Authority was held at the Hotel Rockingham in Portsmouth, New Hampshire. The full Board was present and consisted of Rep. William H. Hinman, Skowhegan, Maine; Attorney Hollis B. Cole, Kittery, Maine; Paul C. Thurston, Bethel, Maine; Frederic E. Everett, Concord, New Hampshire; Dr. James J. Powers, Manchester, New Hampshire; and Frank E. Brooks, Portsmouth, New Hampshire. At that meeting, Frederic E. Everett was elected Chairman of the Authority; William H. Hinman, Vice-Chairman; Hollis B. Cole, Treasurer; and Frank E. Brooks, Clerk. Also present were Eugene E. Reid of Concord, N.H., State Director of the New Hampshire National Emergency Council; Harold J. Lockwood of Concord, Director of the Public Works Administration of Maine, New Hampshire and Vermont; Lucius D. Barrows, Chief Engineer of the Maine Highway Commission; and Daniel H. Dickinson, Chief Engineer of the New Hampshire State Highway Department. The main business of this meeting was the passing of a resolution which authorized the chairman "to file an application to the Public Works Administration to aid in financing the construction of an interstate bridge...." Implicitly, this recognized that federal help would be necessary to pay for the construction. That the government bureaucracy, created by the administration of President Franklin Delano Roosevelt, was already in place to provide such help, is suggested by the attendance of federal program officials (Eugene E. Reid and Harold J. Lockwood) at this meeting.[9]

The second meeting of the Authority was held at the Eastland Hotel; Portland, Maine, on June 16, 1937. Discussion focused on the following items: The offer from the B.& M. Railroad to pay $35,000 per year in rent and the content of the legislation that Sen. Fred H.

Brown of New Hampshire would submit to Congress concerning the bonds to finance the bridge. Meanwhile, Judge Cole was in Washington lobbying Congress and federal officials to approve the application of the Authority for Public Works Administration funds.[10]

At a meeting of the Authority held on October 8, 1937, the engineers bidding on the design of the bridge presented their proposals. These included Parsons, Klapp, Brinckerhoff and Douglass; Fay, Spofford and Thorndike; Moran, Proctor & Freeman; J.R. Worcester & Company; and Harrington & Cortelyou; absent were Robinson & Steinman, who had been invited to bid. The Authority selected Harrington & Cortelyou of Kansas City, Missouri, because theirs was the lowest bid and they were well qualified to do the work, having had extensive experience in designing vertical lift bridges.[11]

The Authority's next task was to secure financing for the project, an endeavor which occupied much of 1938. At the Authority meeting of February 19, 1938, Harrington & Cortelyou reported the general interest of New York bankers in a bond issue paying about four percent interest, if a traffic count proved to be satisfactory; the engineers also suggested that the bridge could be built for $3,000,000 or less including roads.[12] George E. Abbott, Manager of Brown Harriman and Company of Boston, subsequently negotiated with the Authority concerning the possibility of underwriting the bond issue.

Brown Harriman suggested that a traffic and revenue survey be performed, and recommended the firm of Coverdale and Colpitts for this service. In a letter to the Authority dated July 12, 1938 (and attached to the minutes of the Authority meeting of the same date), the manager of Brown Harriman proposed that the firm purchase approximately $1,700,000 Thirty-year 3-3/4 percent Sinking Fund Bonds from the Authority at a price of 97-1/2. In addition to the usual conditions including the maintenance of a satisfactory market, this understanding was subject to the receipt of a satisfactory traffic report from Coverdale and Colpitts, the receipt of a portion of the funds under the proposed Public Works Administration grant, and the simultaneous execution of a satisfactory rental agreement with the Boston and Maine Railroad.

As suggested, Coverdale and Colpitts prepared a report estimating earnings and operating expenses of the proposed bridge. Based on traffic counts made in August 1938, the firm estimated annual traffic for 1938 on the Memorial Bridge of 4,800,000 vehicles, and in 1940 of 5,280,000 vehicles; it was estimated that the Interstate Bridge would attract 15 percent of that traffic, or 792,000 vehicles. Average receipts for vehicles for the first five-year period of operation of the new bridge were estimated at $91,500, based on declining traffic growth rates after the first year of 3, 2-1/2, 2 and 1-1/2 percent, respectively. Annual operating expenses of the Authority were estimated to be as low as $27,000, or as high as $50,000, depending on whether it had to maintain its own administrative and clerical staff, and on whether the two states assumed maintenance of the highway portions of the bypass. On the basis of the report, Brown Harriman Company withdrew from the project. The Authority then authorized Dr. Powers to apply to the Reconstruction Finance Corporation, a federally-funded entity, for the necessary financing.[13]

Ultimately, financing of the Interstate Bridge was obtained through a grant of $1,420,040 from the Public Works Administration, a loan of $1,769,000 from the Reconstruction Finance Corporation secured by 4 percent bonds (amounting to 45 percent and 55 percent, respectively, of the total), and by the sale of an additional $320,000 4 percent bonds to complete construction. Outstanding bonds of the initial issue were redeemed in 1945 from the proceeds of $2,400,000 Bridge Revenue 2-1/4 percent Refunding Bonds dated June 1, 1945. In 1955, a new issue of $2,100,000 2.10 percent bonds was carried out, underwritten by Tripp and Company, "...to provide funds for the redemption of all outstanding Bridge Revenue 2-1/4 percent Refunding Bonds of the Authority and to finance needed improvements to the bridge and its approaches..."[14]

At meetings of November 1 and 3, 1938, the Bridge Authority passed resolutions accepting the grant from the Federal Public Works Administration and the loan from the Reconstruction Finance Corporation, set out procedures to be followed and appointed attorneys and bond counsel. The bond counsel chosen was Story, Thorndike, Palmer and Dodge of Boston, Massachusetts, with Arthur E. Sewall of Portsmouth and McLane, Davis and Carlton of Manchester, New Hampshire, as retained attorneys, along with the firm of Ropes, Gray, Boyden and Perkins of Boston. The National Shawmut Bank of Boston was to "make the arrangements with Ropes, Gray, Boyden and Perkins and advise the Authority and its Counsel, Mr. Sewall." The meeting of November 1, 1938, authorized the employment of "a competent secretary and bookkeeper," who arrived officially on November 3, in the person of Miss Sarah Mildred Long, who thereby began a remarkable career of service with the Bridge Authority which has continued for a half century.[15]

Bids for the construction were advertised by the Authority on December 2, 1938, and awarded at its meeting on December 16. The winning bidders were:

Unit No. 1: Substructure Frederick Snare Corporation 114 Liberty Street New York, New York	$682,076.75
Unit No. 2: Superstructure The Phoenix Bridge Company Phoenixville, Pennsylvania	994,351.00
Unit No. 3: Maine Approach Littleton Construction Company 46 Union Street Littleton, New Hampshire	305,428.75
Unit No. 4: New Hampshire Approach John Iafolla Construction Company 11 Van Brunt Avenue Dedham, Massachusetts	496,013.00
	$2,477,869.50

Also at this meeting, the Public Works Administration grant was noted as being $1,309,090, and the Reconstruction Finance Corporation loan as $1,600,000, for a total of $2,909,090. The excess was allotted to:

Preliminary	$5,000.00
Land: Maine	30,000.00
Land: New Hampshire	129,000.00
Contingencies	39,873.50
Engineering @ 6%	148,680.00
Interest during Construction	42,667.00
Legal and Administrative	36,000.00
	$431,220.50

The contracts were intentionally broken up at the outset to allow participation by the smaller contractors typical of upper New England. In addition to Littleton and Iafolla, mentioned above, these contractors included Bancroft-Martin Rolling Mills of South Portland, Maine; Lewiston Welding Company of Lewiston, Maine; and Ellis C. Snodgrass of Portland, paving contractor for the bridge.[16]

III *The Design*

RECOGNIZED AS AN extraordinary engineering project when it was being built, the Interstate Bridge, when completed, became one of the dominant architectural monuments of the Seacoast area. The two towers can be seen from almost anywhere in Portsmouth; they display an architectural treatment which is just now being recognized for its high quality. The connected Route 1 Bypass, if not perfect in the light of later developments, has proved amazingly adaptable to changing needs over a period of fifty years. A word about the history of vertical lift bridges in the United States will help to put the Interstate Bridge in perspective. The first major vertical lift bridge in the United States, the South Halstead Street Bridge in Chicago, was designed by J.A.L. Waddell in 1892. From 1907-1914, Waddell was in partnership with John Lyle Harrington, later one of the founders of Harrington and Cortelyou. The two became so commonly identified with the typical early vertical lift bridge that it is frequently designated a Waddell-Harrington bridge.[17]

Most of the early vertical lift bridges were designed as purely utilitarian structures. The Memorial Bridge (1923), located downstream from the Interstate Bridge and also running between Portsmouth and Kittery, is characteristic of the period. Designed by J.A.L. Waddell, it was built by the American Bridge Company. The lift span is a 297'-1" Warren truss, which is flanked by Warren trusses of equal length, but with arched top chords. The three trusses were floated into place on barges, towed by tugs, with the ebbing tide used to set them onto the piers.[18] The backs of the four-pier towers slope up and in, in the traditional way. No attempt is made to hide the bracing

Memorial Bridge construction. Floating one of the lateral spans into place. Portsmouth shoreline behind, with the Moffatt-Ladd House above the pilothouse of the left-most tugboat. The cranes are at coal pockets on Nobles Island. C.1922. Photographer facing south from Badger's Island, Kittery. Photograph courtesy of Captain Norman Y. Holt.

Memorial Bridge construction. Staging on the towers. C.1922. Ferry and Badger's Island landing in foreground. Photographer facing southeast. Photograph courtesy of Captain Norman Y. Holt.

Interstate Bridge, C.1970. Photographer facing northwest. Photograph by Douglas Armsden.

of the towers or to give them any form other than that dictated by engineering considerations. Although the Memorial Bridge has very fine sculpture over its south portal in New Hampshire, connected with its World War I memorial dedication (in addition to an outstanding relief sculpture by Bashka Paeff in the Kittery, Maine, park on axis with the bridge), it remains a purely utilitarian structure and, as such, typical of its era.

By the 1930s, bridge engineers were beginning to relate bridge design to contemporary architectural currents. While Otis Ellis Hovey had complained, in 1926, that "the great height of the towers detracts from the architectural effect," the following decade saw vertical lift bridge towers become the central feature of the design.[19]

The Interstate Bridge further modified the traditional arrangement of four columns in each tower, used by all large vertical lift bridges. While using rectangular, four-column towers (as opposed to those, like the Memorial Bridge, in which the side away from the lift span slopes up to meet the other side at the top), it narrowed the lateral sides still further than some recent bridges and covered them with a welded shell, flaring at the bottom where it meets the pier, and ending in a low pediment or gable at the top.

The Interstate Bridge was featured in *Movable and Long-Span Bridges* by Hool and Kinne, which was revised (in 1943) shortly after its construction by R.R. Zipprodt and H.E. Langley, the latter being Bridge Engineer of the New Hampshire State Highway Department.[20]

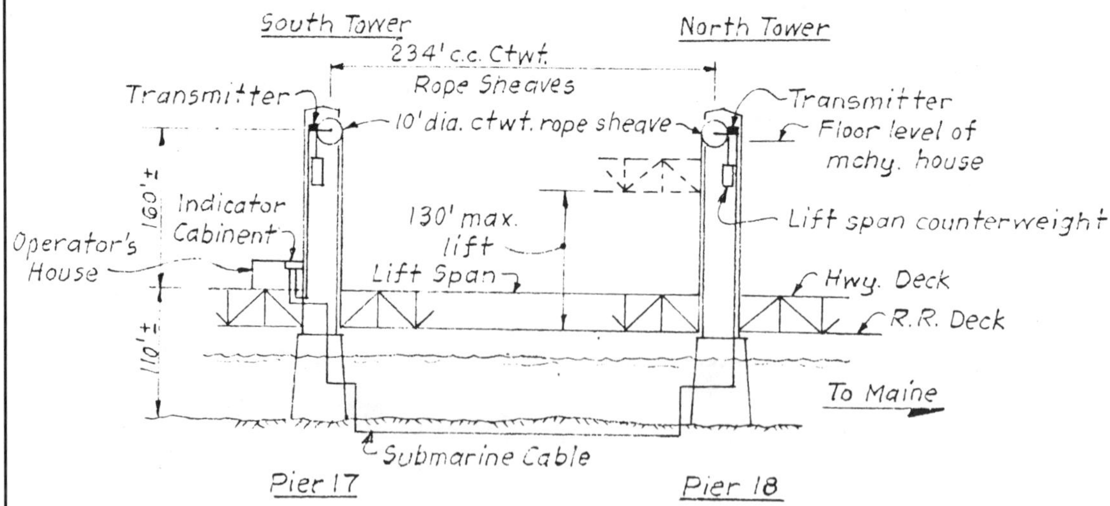

OPERATIONAL NOTES:

1. Span will automatically stop when out of level 5" between its N & S ends.
2. To restart, set controls as reqd. to level & hold in by pass button until green level lt. operates. Then resume normal operation.
3. Green level lt. indicates span within 1/2" of level.
   Red south up lt. indicates south end 5" higher than north end.
   Red north up lt. indicates north end 5" higher than south end.
4. Following an automatic stop due to the span being out of level, and until green level lt. operates, span controls will be inoperative to:
   1. raise both ends simultaneously.
   2. lower both ends simultaneously.
   3. raise up end.
   4. lower down end.

Diagram of Interstate Bridge lift span controls and operating instructions. Harrington and Cortelyou, Consulting Engineers, Dec. 1971.

As with virtually all bridges of this type, the main feature is the lift span between the two towers; the span is raised for the passage of marine traffic by lowering the connected counterweights, which are suspended by cables from each tower. In the fully open position, the counterweights are lowered almost to the bridge highway deck; with the span closed, however, they rest at the tops of the towers, just below the machinery rooms. The span can be stopped at any point between "full open" and "closed," thereby reducing the amount of time that highway traffic is interrupted when a small vessel goes through (obviously, a lift of 40' requires less time than a "full open" lift of 135', for example). Except in the very rare cases where a vertical lift bridge is operated by remote control, it requires an operators' house to control the lifts; in this case, it is cantilevered out over the water from one side of the highway deck, but the location varies from one bridge to another, depending on a number of factors. In the Memorial Bridge, the operators' house is located above the roadway, in the center of the lift span.

In a few cases, where the body of water is narrow, the only span is the lift span (this was true of Waddell's first bridge in Chicago), but most vertical lift bridges have a series of trussed spans leading from the shore to either end of the lift span, as found here. By "trussed spans" is meant, of course, metal spans stiffened and strengthened by vertical and diagonal members. In the nineteenth century, numerous truss systems were developed, defined by the differing placement of elements and whether they were under tension or compression; these are known by various names (often commemorating their inventors) such as Pratt, Howe and Warren. These trusses rest on concrete piers, which in turn rest on bedrock, often some distance below the riverbed ("pier" in this sense is not to be confused with the vertical steel equivalents of posts or columns in the superstructure of the towers).

In engineering and construction, the superstructure (basically, everything from the top of the piers upward) and the substructure (everything from that point downward) require different skills, so that the construction, at least, is often carried out by different contractors. Moreover, the U.S. Route 1 Bypass, of which the Interstate Bridge is a part, was purely a highway project, so that companies specializing in that field contracted to do the work. The Maine-New Hampshire Interstate Bridge (the Sarah Mildred Long Bridge) runs approximately southwest (Portsmouth, N.H.) to northeast (Kittery, Maine); for the sake of convenience, the ends shall be called 'south' and 'north', respectively. Of the connected bypass, approximately 2-1/2 miles are located in New Hampshire, and 1-1/2 miles in Maine, exclusive of the bridge. The total length of the project is 4.4 miles.

The bridge has a total length of 2798'—4-1/2", center- to-center of bearings at the north and south abutments. The bridge proper consists of five Warren trusses (the basic Warren truss has diagonals which are alternately placed in tension or compression; thus, it is easily recognized by its triangular outline, but most have vertical stiffening members as well)—the lift span of 224', flanked by spans of 226' which are, in turn, flanked by spans of 227'.

Thus, the Interstate Bridge consists of the central lift span, flanked by truss spans. To the north and south of the latter are series of plain steel girder spans (i.e., not trusses, but simply steel beams rivetted together, in outline similar to most highway overpasses seen today), seven to the north, totalling 546' 6" in length, and fifteen to the south, totalling 1083' 10-1/2" in length. The truss part of the bridge is symmetrical in elevation, and in plan the bridge proper is straight from bank to bank. However, unlike the shorter Maine portion, which is straight from embankment to river, the relatively long New Hampshire approach makes a shallow, sweeping curve between the two, crossing Portsmouth's North Millpond in the process. Thus, in elevation, the bridge provides a subtle balance of verticals and horizontals (the horizontal of the highway deck with the verticals of the two towers), and in plan balances the symmetry of the straight central portion with the assymmetry of the long, curving south approach.

The tower height from the top of the concrete pier to the tops of the side parapets (flanking the roof) is 207'-6", with a total height above mean high water of 218'. The top of the pier has two steps, giving it the visual as well as actual, function of a base. The sides have a batter (i.e., slope inward as they rise) of 1:200 and the front and back a batter of 1:120. In other words, the whole tower tapers from the flaring base to the low-arched roof.[21]

Other subtleties in the towers are the twin pilaster strips (a pilaster being a feature meant to suggest a column, but flat and attached to a wall, rather than round and freestanding) on the east and west sides, and the indentations at the tops of these sides which stop the eye at the low-gabled parapet forming the end of the tower roof. Architecturally, the style is related to Art Deco and Modernistic buildings of the 1920s and 1930s, particularly those skyscrapers which used abbreviated classical forms, such as pilasters, to emphasize the verticality of the structure. The stress on verticality, with classical overtones, was further carried out in the treatment of the machinery house windows (the machinery houses forming the closed-in tops of the towers), placed in vertical line with the taller openings in the counterweight housing below them. This constitutes an unusual example of the Modernistic style applied to a bridge.

Mechanically, the vertical lift bridge is intriguing by comparison with other bridge types. The lift span of the Interstate Bridge weighs approximately 1500 tons (as built—1,788,000 lbs.; design—1,880,000 lbs.) and is balanced by two counterweights of roughly 750 tons each, which can be adjusted for new pavement or other weight changes by moving, adding or subtracting numerous concrete weights contained within them. The lift span is operated by one General Electric Motor, type MTC, of 100 hp. @ 865 r.p.m., 3 phase, 60

Interstate Bridge. Railroad deck. September 4, 1987. Photographer facing south. Photograph by author.

cycle, 440 volts, located at the top of each tower. This runs through a single reduction unit located next to it, at a ratio of 43.7 to 1 (simply put, the reduction gear converts speed to power, much as an automobile is shifted into low gear). In addition, the span is levelled electrically by a synchronous-tie between two General Electric Motors, type MR, 50 hp. @ 900 r.p.m., 3 phase, 60 cycle, 440 volts, one on each tower.

This electrically-synchronized method of levelling the span differs from the mechanically-operated system employed on most older lift bridges such as the nearby Memorial Bridge (opened 1923). It eliminates the need for separate uphaul and downhaul cables which wind onto or off of drums located at the tower tops or the middle of the lift span; for that reason, cable life is extended considerably. On the other hand, the use of electrical means of levelling the span make it somewhat more prone to going out of level ("out of skew"), particularly when high winds, heavy snow loads and freezing conditions are encountered—this may be partly a function of age and mechanical wear, however.

Thus, the lift system here is 'free-running', as opposed to the older, mechanically continuous system. Essentially, the free-running system consists of cables run from the lift span over the drums at the tops of the towers to the counterweights; the electric motors move the drums, hence the cables, in turn moving the counterweights and the lift span (The drums are 10' in diameter, and the counterweight ropes, which lead from each upper corner of the lift span, consist of sixteen 1-3/4" wire-on-fiber core, 6x19 ropes). In effect, the electric motors merely start the process of raising or lowering the lift span, after which the weight (of span or counterweights) does most of the work. The older type of vertical lift bridge, represented by the Memorial Bridge between Portsmouth and Kittery, has a mechanically continuous system; this consists of two sets of cables, known as uphauls and downhauls, which simultaneously wind on and off the drums to raise or lower the span. In the Memorial Bridge, these drums are located in the operators' house; they are in addition to the drums which lead the

Interstate Bridge. South machinery room. South downstream bull wheel (diameter 10') behind Hiram Young, Maintenance Superintendent. Mr. Young holds a walky-talky to communicate with the bridge operator. September 4, 1987. Photographer facing northeast. Photograph by author.

Interstate Bridge. South machinery room. Left to right: Brake, lift motor, reduction gear, levelling motor. September 4, 1987. Photographer facing southwest. Photograph by author.

cables over the tower tops to the counterweights.

In the Interstate Bridge, each tower has seven accelerator contactors, vertically arranged, to pull resistance out of the lifting motor as the span moves. As placed in the machinery rooms, the reduction gear is the large, centrally-placed motor with the lifting motor on one side, connected to a brake, and the levelling motor on the other side. In addition, there is a Selsyn Transmitter Motor in each tower, connected by a chain to the main drum, to measure the revolutions of the drums in each tower, thereby telling the operator if the span is level. The original reduction gears were made by Earle Gear; these were replaced with D.O. James reduction gears in 1976. The new gears have the same gear reductions, but single driving gears instead of the former double ones, and ball-and-roller bearings instead of sleeve bearings. From the exterior, one can easily see where a panel was

Interstate Bridge. South counterweight. Concrete weights inside the steel shell. These can be moved, added or subtracted to compensate for repaving or other factors affecting the weight and balance of the lift span. September 4, 1987. Photographer facing west. Photograph by author.

removed from the machinery room on each tower to replace the reduction gears (obviously the weld was not "ground smooth" as specified in the original plans).[22]

The Interstate Bridge was designed by Harrington & Cortelyou, Consulting Engineers, of Kansas City, Missouri, the principal designers of the bridge being Frank M. Cortelyou Sr. and Theodore J. Cambern. Mr. Cambern was the chief draftsman of the firm in the late 1930s and may have been responsible for the architectural design; he was the father-in-law of Willis Grinstead, presently a partner in the firm. John Lyle Harrington (known as "Colonel Harrington") visited the job site frequently; by the time the bridge was built, he was one of the older generation of bridge engineers. Richard Tatlow was the firm's Washington representative; he was instrumental in getting the project approved by the federal government. Unlike Colonel Harrington, who was advanced in years, H. Denton Peoples, who served as the firm's resident engineer from the fall of 1939 until the completion of the project, was in his early 30s; he was responsible for seeing that the bridge was constructed as engineered by Harrington and Cortelyou, and was assisted by F. Norvell ("Mike") Larkin and Edwin M. Newman (9). Jerry Tripp, of Tripp & Company in New York, devoted much effort to working out the financial arrangements, according to Miss Long.[23]

Edwin M. Newman, who served as Assistant Engineer for Harrington and Cortelyou on the project, reported that most of the lift spans designed by the firm were single-purpose structures, carrying a railroad or highway, but not both. Therefore, the Interstate Bridge was unusual for them in having a highway on the top chord (i.e., the top plane of the trussed spans) and a railroad on the bottom. Because of this arrangement, neither the motors nor the operators' house could be placed within the trusses (as they were on the Memorial Bridge—a single-purpose highway bridge with the road on the lower chord), but had to be placed elsewhere.[24]

After the bridge was completed, Harrington and Cortelyou continued to serve as consulting engineers to the Bridge Authority; in 1965, the firm designed a moveable small craft span to decrease the number of lifts of the main span. The new span is the southernmost plain girder span on the north (Kittery) end of the bridge; part of the railroad deck, this 85' span was modified to lift and retract inside the adjacent truss span to the south, without affecting the highway deck above. The purpose is to provide an alternative opening for small craft so that they can pass through the railroad deck without disturbing the flow of highway traffic (which happens when the main lift span is raised). Although the retractable span is a very old type of bridge, its use in this way may be unique among American vertical lift bridges. The new construction was carried out between 1966 and 1968. By that time, the Eastern Branch of the Boston and Maine Railroad had been abandoned, and rail traffic over the bridge was limited to that going to the Portsmouth Navy Yard; therefore, the small craft span is ordinarily left open in the summer months, when most small craft traffic occurs. It is operated from a platform between the two decks.

Of interest is a proposed major rebuilding of the Interstate Bridge, outlined in a report dated March 15, 1984. Unbuilt, this was entitled "Portsmouth Harbor and Piscataqua River Navigation Improvement Project; Alteration of Maine-New Hampshire Interstate Bridge...." The study was carried out under the auspices of the U.S.Army Corps of Engineers, with Cleverdon, Varney & Pike, Inc., and Harrington & Cortelyou, Inc., Consulting Engineers. The Navigation Improvement Project was authorized by Resolution of April 23, 1970, by the U.S. Senate Committee on Public Works.

The report examines alternatives to increase the horizontal clearance of the navigation channel from 200 to approximately 300 feet. Parenthetically, it should be noted that the bridge as built sits at an angle to the channel, due to the demands of land connections, which narrows the useful channel considerably—probably 30' or 40'. The project is worthy of study because it shows how the same engineers would redesign the bridge with a longer lift span today. Alternatives considered included building an entirely new bridge on a different alignment; rebuilding the existing north tower in place and extending the span southward, building a new south tower; or building two new towers equidistant from the existing lift span to accomodate the new, longer span. The last alternative was recommended, both for economic and aesthetic reasons. Building a new bridge would cost about $50,000,000, while using one existing tower would not save any significant amount of money and would alter span lengths in a way which would unbalance the overall visual composition of the bridge. Its high visibility from the surrounding area, and from I-95 bridge in particular, was noted. The estimated cost of altering the bridge using the last alternative (building new towers and installing a longer lift span) was $15,930,000.

The proposed construction methods included building the new main piers with the use of pneumatic caissons (as were used originally), fabricated in 5 to 10-foot sections so they could be sunk under the existing spans, building the new towers before the old ones were demolished, and fabricating the new lift span on shore so that it could be floated into place on the piers (as was done with the Memorial Bridge). The new towers would be approximately 10 feet higher than the old ones, and 5 feet wider, center to center of columns, "to permit erection outside the present structure." The benefit of keeping the retractable span in operation during construction was emphasized.

In the final paragraph of the report, the engineers reiterated the importance of "preserving the favorable aesthetics of the bridge after alteration." Elevations of the proposed new towers show a marked similarity to the old ones, with the difference that the solid side covering is given a series of five openings, one above the other, apparently surmounted by a low 3-light window at the end of the machinery room. No front elevation of a tower is given, but one can surmise that the fronts and backs would be treated similarly to those of the present towers.

According to Captain Richard Holt, the above bridge alterations were never seriously considered, but had to be reviewed as part of the Piscataqua River Navigation Improvement Project. Nevertheless, it is interesting to see how the same engineering firm would have redesigned the bridge some fifty years later.

IV *The Construction*

THE ENGINEERS OF THE Interstate Bridge faced many challenges in the design phase, due to the haste with which the preliminary plans had to be carried out; this necessitated using incomplete data such as the 1915 railroad survey of the river bottom. As a result, they made major changes in the bridge plans shortly before construction was to begin. These changes had to be approved by the War Department, which was required to rule on any construction which might affect the use of navigable waters (In fact, the involvement of the War Department with vertical lift bridges went back to the 1890s, and it was then that the 135' clearance requirement for the spans became standard). The changes are revealed in a notice sent out by Captain W.S. Moore, Corps of Engineers, for the War Department on October 29, 1938; it is excerpted as follows:

> The modified plans show the proposed bridge to be located 60 feet downstream from and parallel with the center line of track on the present railroad bridge, or a distance of 110 feet downstream from the original approved site; also, the approach on the Portsmouth, N.H. end is to be shortened 64 feet by substituting two 88-foot spans for three 80-foot spans, and the approach on the Kittery, Maine end is to be lengthened 74 feet by adding one 50-foot span and substituting three 88-foot spans for three 80-foot spans. The net result will be to move the lift span 34 feet toward the Portsmouth shore and toward the deeper section of the channel. The clearances of the lift span are to remain as shown on the original plans; that is, hor-

Interstate Bridge construction. Pier in North Millpond, Portsmouth. 1939. Photographer facing west. Photograph by H.D. Peoples.

> izontal clearance 200 feet, vertical clearance 135 feet at mean high water, and 142.8 feet at mean low water.
>
> These changes are desired in order to simplify the highway and railroad connections at the ends of the bridge and to adapt it to topographic conditions only recently revealed by an accurate check survey made under the directions of the consulting engineers employed by the applicant.

The proposed changes were approved by the War Department, and by the end of February 1939, bridge construction was ready to begin. Two tractor type cranes immediately set to work repairing the docks on Nobles Island in Portsmouth; these docks would serve as moorings for the barges and as an unloading wharf for the contractor during construction. A diver was hired to inspect the pilings of the docks, which had not been used for many years.[25]

The B.& M. Railroad was still using the old bridge, at 119 years reported to be the oldest pile bridge in this part of the country. It received its final inspection by a diver in April, about the time the work on the new bridge project went into full swing. The project of hauling fill for the Route 1 Bypass in Kittery was delayed by spring "mud season", but had resumed by April 20, 1939 (the State had temporarily banned heavy hauling on state highways). Foundation work had begun on the Eliot Road underpass for the Kittery Bypass. On the Portsmouth Bypass, the B. & M. Railroad tracks were being relocated

(Left) Interstate Bridge construction. South approach. Old railroad bridge to right. C.1939. Photographer facing south. Photograph by H.D. Peoples. (Above) Interstate Bridge construction. North Mill pond, Portsmouth. C.1939. Photographer facing southwest. Photograph by H.D. Peoples.

Interstate Bridge construction. Photographer facing north. C.1939. Snare Corp. concrete plant on Nobles Island to right. Photograph by H.D. Peoples.

Interstate Bridge construction. Photographer facing east toward Portsmouth. C.1939. Photograph by H.D. Peoples.

by the John Iafolla Company in the vicinity of the Morley Button Company, with the Concord branch also being relocated to that route, to use a single overpass as mentioned above. James Gardiner, the federal Works Projects Administration engineer on the project, reported 155 men on the payroll.[26]

By June 15, 1939, with approximately 300 men working on the bridge, rough grading on the Maine approach was nearly complete along with the Eliot Road underpass and the B.& M. Railroad tunnel, while concrete was being poured for the Middle Road overpass on the Portsmouth side of the river.[27]

Interstate Bridge construction. Snare tugs and derrick boat. Memorial Bridge in background. C.1939. Photographer facing east. Photograph by H.D. Peoples.

Interstate Bridge construction. Caissons at dock on Nobles Island, Portsmouth. Consolidation Coal Co. facilities to right rear. C.1939. Photographer facing east. Photograph by H.D. Peoples.

A month later, the first of nine pneumatic caissons ("pneumatic" because they depended on the use of compressed air for the working environment) had been placed, and the *Portsmouth Herald* reported that "Further sinking of the caisson will be performed by

Interstate Bridge construction. Caisson #19 enroute from Nobles Island to the bridge. C.1939. Photographer facing east. Photograph by H.D. Peoples.

Interstate Bridge construction. Steelwork headed for Maine. November 15, 1939. Photographer facing north. Photograph by H.D. Peoples.

men, commonly known as sandhogs, working under compressed air, excavating beneath the caisson allowing it to sink through the overlying stream bed to the ledge below." The caissons were prefabricated steel boxes which were towed out to the construction site from Nobles Island, sunk to the river bottom where the piers were to be, and

Interstate Bridge construction. Nobles Island, Portsmouth, in background. November 15, 1939. Photographer facing south. Photograph by H. D. Peoples.

pumped out. The sandhogs then entered the caissons through airlocks (which can be seen projecting from the tops of the caissons in photographs) and proceeded to excavate beneath the caisson until it could sink through the bottom earth and debris to settle on bedrock. At that point, the bedrock was thoroughly cleaned off and the caisson was sealed to the bedrock with concrete. Concrete was then poured into forms within the caissons, a short distance inside the walls, to build up the pier. All of this work was done with air being pumped into the caisson to maintain a constant air pressure. This pressure kept water from seeping into the caisson from the river. The sandhogs experienced the same conditions (minus the enveloping water) as

Interstate Bridge construction. House for sandhogs to change clothes, and decompression chamber for their use when needed. C.1939. Photograph by H.D. Peoples.

Interstate Bridge construction. Pneumatic caisson in place, and one being towed. November 15, 1939. Left and right arrows indicate supply shafts, center arro w indicates a man shaft. Photographer facing east. Photograph by H. D. Peoples.

divers, being required to emerge gradually from the caisson when their shifts were done so that they would not get the bends (a crippling and potentially fatal condition where nitrogen bubbles form in the blood). A hospital airlock was located on Nobles Island for the treatment of sandhogs who got the bends (they were put back under pressure and then very gradually decompressed—i.e., the atmosphere

Interstate Bridge construction. Concrete plant on Nobles Island, Portsmouth, with new bridge behind. Convenient rail connection! November 15, 1939. Photographer facing west. Photograph by H.D. Peoples.

was returned to normal). The work was both novel and dangerous, it being said that the sandhogs in the caisson could reach out and wash their hands with the river water which was held back by the air pressure. The sandhogs, incidentally, were highly paid for the time, receiving fourteen dollars a day for about one hour of work. One of the sandhogs is believed by Mr. Peoples to have perished from the bends, but the author has been unable to discover his identity. The greatest depth reached would be 87.02' below mean sea level at Pier 16, the one next south from the south tower of the bridge.

Interstate Bridge construction. Land piers. Nobles Island; Portsmouth, N.H., in fore - ground; it served as the construction base. The walkway leads over the North Millpond to the mainland. The Route 1 Bypass route cleared behind the crane. 1939. Photographer facing south. Photograph by H.D. Peoples, Resident Engineer.

The concrete for the bridge piers was made in a concrete plant on Noble's Island and pumped through 6" pipe out to the bridge and down through supply shafts into the caissons. This was known as the "Pumpcrete" system, meaning "pump concrete," and it is believed by Denton Peoples to have been the first time this was used to seal a pier inside a pneumatic caisson. The caissons had three shafts—two supply shafts, each about three feet in diameter, and a man shaft to provide access for the workers. The Pumpcrete system worked well and saved an immense amount of time over the old system, which was to "bucket" the concrete into the caisson. Within the caisson, the air pressure was maintained at approximately fifty pounds per square inch, which was about the maximum pressure to which workers could be subjected.[28]

The caissons were anchored in place by a series of fifty-ton anchors attached with cables, and it is that fact which figures in the railroad accident of September 10, 1939, off the wooden pile bridge just upstream from the construction site. The accident, mentioned elsewhere, occurred in the evening, and involved a passenger train southbound from Portland, Maine. The engine and tender fell through one of the spans, an event not witnessed by any survivor. Subsequently, the nearby Caisson 17 was found floating against the old bridge, obviously having been involved in some way in the accident. As Mr. Peoples relates, "...that was a question of the hen and the egg, who

Interstate Bridge construction. Land piers on Nobles Island, Portsmouth. Old railroad bridge at upper left. 1939. Photographer facing north. Photograph by H.D. Peoples.

comes first?" The contractor claimed that the engine and tender struck the anchor cables (which led through the space below the span) on their way down, causing the caisson to move toward the pile bridge, while the railroad asserted that the cables had moved first, weakening the structure and causing the accident. The engine and tender ended up on the river bottom between Piers 17 and 18 of the Interstate Bridge, where they remained until c.1966; at that time, they were moved because they were creating a shoal. Piers 17 and 18 are the ones flanking the lift span, 17 being the south pier.

The concrete piers, which were subject to deterioration n the area alternately covered and exposed by the tides, had their upper parts faced with granite ashlar masonry, the joints being filled with lead wool to keep the water out.

Scottish masons, highly-skilled at their craft, were brought in by the Snare Corporation to do this masonry. Unlike the deepwater piers, the land piers and those in shallow water could be built under the normal atmosphere; they were "seat pile cofferdam construction," open above, to be dug out with a steamshovel clamshell and by hand.

Several different kinds of cranes were used for the bridge construction, including crawler cranes for the land work on the approaches, barge derricks on the river for setting up the false work and support used to construct the bridge superstructure, and a travelling crane which moved along on the bridge deck from pier to pier as the structure advanced into the river. Contemporary photographs reveal the use of steam power in all of this machinery—tugs, cranes, and the trains which brought material (including coal) directly to the land base on Nobles Island.

Construction methods were somewhat different than today, specifically in the use of rivets throughout the bridge superstructure (today, much of this would be welded or bolted). A lobster boat was kept on standby in the river to pick up steelworkers who fell from the structure and nobody was lost in that way. Throughout the project, the old pile bridge was used for a construction platform, with the part in the middle of the river remaining in place until sometime after the dedication of the Interstate Bridge. However, after the train accident, it was never again used by the railroad. Photographs of the moved caisson taken after the accident clearly illustrate the strong tidal currents of the Piscataqua River which certainly added to the difficulties of the project.

In addition to the delay caused by the accident, which is difficult to gauge at this point, considerable delays were caused by the weather in the winter of 1939-1940. Storms and high winds meant that the steelworkers could only work four out of every five days during this period.[29] These delays caused serious concern because the Public Works Administration grant would not cover work done after June 23, 1940. The Authority asked for an extension of the date for final completion and the P.W.A.'s sharing in final costs to October 15, 1940; the reasons given were "conditions beyond our control" and "extra work not anticipated at the time of awarding the contracts." The extension was granted.[30]

As the project neared conclusion, the operation of the lift span had to be tested. Denton Peoples, who was Resident Engineer, vividly remembered this operation:

> Now there always (is) a question of whether all this, relays and everything, is going to work proper. Well, I think the specifications said the bridge could stand to be six inches out of level from one end to the other. Well, they had an electrical subcontractor doing the work...for Phoenix, and their foreman then started to raise this bridge...Of course, both ends are supposed to raise together and down together and so forth. Well, that six inches was not enough. They got it way out of level before they hollared back and forth to stop it. Well, what Phoenix did then, they put her down on the piers, pressin' it on their base, and then they called (General Electric), and I think (General Electric) furnished all that gear... The motors and the controls and everything, and they sent an expert out and spent several days goin' over all the circuits, the lines and checkin'

em out and the guy knew what he was doin' and finally, after several days of checking and changing and altering and replacing, he said, 'Well, today's the day...Let's go up with it.' They started up and everybody was pretty antsy about whether this was goin' to work or not...get it out of level, they could lose that whole span. Well, it worked fine. They took it up slow and went up, then we got to the top, then we got down, brought it down, and then after that we just kept doin' that several times. Everybody got real positive thinking that it was going to work.

The project encountered the unexpected, both in the construction of the bridge proper and in the building of the bypass. One incident concerned the dog cemetery of Dr. Haven T. Paul, located generally between the present Holiday Inn and the northbound exit onto Woodbury Avenue. This was the subject of a wonderful article (with photo) in the *Boston Herald* for May 19, 1940, entitled "Millions of Motorists to 'Turn Out' for Dog, Dead 20 Years." In fact, the course of the road was changed slightly to avoid the necessity of notifying 100 dog owners all over the country; "'When the owner of Bonnie Belle, for instance, picks up her newspaper in North Brookfield, Mass., and discovers that her dog's grave has been changed, I won't assume the responsibility,' explained amiable Dr. Paul to the engineers' attorneys, Messrs. Sewall and Carleton." Another (human) cemetery caused problems because bones were found buried outside the fence; explained as slaves because of the location, they were reinterred, as reported by the *Portsmouth Herald* for November 9, 1940.

Contrary to expectations, the Bypass has never been the express highway envisioned by planners and local promoters. This is due to the large number of businesses, mainly gas stations, which were immediately attracted to the road and have remained there; vehicles using these facilities slow traffic as they leave and enter the bypass. Even before the bridge was opened, officials (somewhat belatedly recognizing the potential for congestion) were fighting the 'gas station wars'. On August 5, 1940, it was reported that Mayor Kennard E. Goldsmith had carried through on his promise to veto any more gas station applications by turning down the Portsmouth City Council's approval of Roger E. Young's petition to erect a station on Lafayette Road "near the junction of that road and the highway now being built in connection with the new bridge..."[31] The veto would have little effect because the Bridge Authority had control of the bypass itself (Lafayette Road is south of it and under the City's jurisdiction) and it was legally unable to prevent access as long as it was done according to state specifications. The whole concept of limited access highways was a new one at the time, so the necessary legislation was not in

Interstate Bridge dedication. Afternoon of November 8, 1940. Part of old Railroad Bridge to left. Photographer facing north from south tower. Photograph by E.M. Newman.

Interstate Bridge. Three engineers (left to right): H.D. Peoples, F. Norvell Larkin, John Lyle Harrington. November 8, 1940. Photograph by E.M. Newman, a fourth engineer.

place. Among the earliest businesses to locate on the bypass was one very closely identified with that era, being one of the earliest highway chains—The Howard Johnson's Restaurant opened for business at noon on November 9, 1940, the day after the bridge dedication, advertising a Sunday chicken dinner for 75 cents ("Regularly $1.00").[32]

The bridge was completed in a period when war was expected, and Portsmouth had its first blackout, organized by the Civic Patrol, on October 27, 1940. Nevertheless, the dedication of the new Interstate Bridge at 2 p.m. on Friday, November 8, 1940, was a happy occasion. Governor Francis P. Murphy of New Hampshire, Hollis B. Cole of Kittery (Treasurer of the Bridge Authority) representing Governor Lewis O. Barrows of Maine, and approximately 800 residents of the two states gathered on the bridge, following a luncheon for officials at the Hotel Rockingham in Portsmouth. Two little girls cut a ribbon representing the boundary line of the two states—Deborah Cole, seven-year-old daughter of Mr. and Mrs. Hollis B. Cole and Joan Everett of Concord, New Hampshire, seven-year-old granddaughter of Highway Commissioner Frederic E. Everett, chairman of the Bridge Authority. The two girls received inscribed silver scissors as a memento of the occasion.

Following the ribbon-cutting, a speaking ceremony took place in front of the toll booths, on the New Hampshire side. Governor Francis P. Murphy of New Hampshire was the principal speaker (Governor Lewis O. Barrows of Maine was unable to attend). Other speakers were L.H. Chellman of New York, senior engineer of the Regional Federal Public Works Administration; Rear Admiral John D. Wainwright, Commandant of the Portsmouth Naval Shipyard; Laurence F. Whittemore, assistant to the president of the Boston & Maine Railroad; Mayor Kennard E. Goldsmith of Portsmouth; Carroll Sterling, chairman of the Kittery Board of Selectmen; and Frank M. Cortelyou, Sr. of Harrington & Cortelyou, consulting engineers for the project.

The bridge was opened to traffic at 3:30 p.m., but was toll-free until midnight. The first (dime) toll was paid by Edwin H. Kimball of Manson Avenue in Kittery.[33]

In the excitement of opening the bridge, something was forgotten, as Denton Peoples recalled:

> Well, what they forgot to do was to get a money counter and the first (full) day they opened the bridge, everybody drove across the bridge back and forth...a lot of traffic, and they had a heck of a time tryin' to count the money that night, so the first thing they did was to order some kind of money counter so they wouldn't have to go to that trouble.

The day of the dedication, almost the entire edition of the

Interstate Bridge and U.S. Route 1 Bypass, Portsmouth, N.H. Toll booths with bridge behind. C. November 1940. Photographer facing north. Photograph courtesy of Harrington and Cortelyou, probably by A.L. Belcher.

U.S. Route 1 Bypass; Kittery, Maine. Route 1 diverging behind "Boston" sign. C. November 1940. This cut in the center barrier only existed for about two years; in 1942, a southbound exit on the right side was created, so that motorists could make a loop under the bypass to reach the Kittery Traffic Circle, also opened in that year. The underpass became part of Route 236 when it was built in 1956. Walsh Brothers Moving Co. now occupies the left corner. Photographer facing south. Photograph by A.L. Belcher, courtesy of Harrington and Cortelyou.

Portsmouth Herald (headlined: "MAINE-N.H. INTERSTATE BRIDGE OPENED TO PUBLIC") was devoted to stories and photos concerning the bridge. Subheadings were: "Three-Million-Dollar Structure Dedicated; Toll Free Until Midnight; Will Carry Both Motor, Rail Traffic Over River; Ceremonies Today Mark Successful Culmination Of Huge Project; Will Reduce Congestion in City." The front-page photograph was entitled "Tall Towers of the New Bridge," and below it were facts about the bridge. Photographs depicted the beginning of the project (December 23, 1938; on the Kittery Bypass), various aspects of the bypass and bridge, and advertisements from the many contractors who worked on the project. On the bridge, a bronze plaque facing the roadway on the concrete handrail parapet at the north abutment (west side) commemorates the bridge opening:

Federal Works Agency
Public Works Administration
John M. Carmody
Federal Works Administrator

Franklin D. Roosevelt
President of the United States

Maine-New Hampshire Interstate Bridge
1940

Lewis O. Barrows	Francis P. Murphy
Governor	Governor
State of Maine	State of New Hampshire

Members

Maine	New Hampshire
William H. Hinman	Frederic E. Everett
Hollis B. Cole	Frank E. Brooks
Stillman E. Woodman	James J. Powers, M.D.
Paul D. Thurston	

Engineers
Harrington and Cortelyou

Contractors
Frederick Snare Corporation
The Phoenix Bridge Company
Littleton Construction Company
John Iafolla Construction Company
Edward L. Paterson and Son[34]

V *Maintenance and Modifications*

IN THE WAR YEARS (1941-45), the bridge served as an important rail link to the Portsmouth Navy Yard, which was building submarines for the U.S. Navy, as well as carrying the Eastern Division of the Boston & Maine Railroad between Portsmouth and Kittery. Traffic on the latter consisted mainly of passenger trains, and was brought back after being detoured on the Western Division (through Dover, N.H., and South Berwick, Maine) following the accident off the old bridge in September 1939. Automobile traffic was very light during the war, and sometimes the tolltakers took in only six or seven dollars during a shift. At its meeting of July 7, 1942, the Authority voted to take out War Damage Insurance in the amount of $2,146,497.18, at a cost of $2,576.41 to be paid to the Kennard Agency, Inc. The insurance was to cover damage in the event of enemy action against this vital transportation facility. The bridge was guarded by a unit of 'colored troops', said to have been part of the 1105th. Infantry Battalion. According to witnesses at the time, there were regularly seven guards. The guardhouses were located on the railroad level. The operator and gateman were responsible for security on the highway level.[35]

During 1942-43, the federal government tried to shift the responsibility for guarding the bridge to civilian guards who would be hired by the Bridge Authority, to free up troops for overseas duty. At the Authority's meeting on August 3, 1943, the chairman was authorized "to notify the Army officials that it would be impossible for the Authority to place guards on the Bridge owing to its financial condition due to lack of patronage because of gasoline rationing." Civilian guards were used later in the War, however.[36]

U.S. Route 1 Bypass in Portsmouth, N.H., with I-95 Highway being built from the south. Portsmouth Traffic Circle not yet built. Howard Johnson's Restaurant in original location, southwest of junction of Bypass and Woodbury Avenue (to right of lower end of Bypass curve in picture). C.1949. Photographer facing south. Photograph by Douglas Armsden.

Howard Johnson's Restaurant on Route 1 Bypass in Portsmouth, N.H., about to be moved for construction of the Portsmouth Traffic Circle. The house in the background is on Woodbury Avenue. C.December 29, 1949. Photographer facing north. Photograph by Douglas Armsden.

Throughout all of these events, toll collection was the heart of the bridge operation. It paid for everything, and the fact that it ceased in 1972 should not obscure its function of paying off the bonds, paying the operating and capital expenses of the Authority, and building up a trust fund for continued operation of the bridge. From the beginning, the toll for passenger cars was ten cents; hence, the bridge became known as the Dime Bridge, a name so generally used that it would appear in the deeds transferring the approaches to the two states in 1986.

With such importance attached to toll collection, every aspect of the operation was monitored. Treadles in the pavement counted the axles on vehicles and printed the number in the machines, and the toll collectors kept track as well. Cash was counted upstairs in the toll booths at the ends of shifts and otherwise controlled at every step of the way. Collection was all manual, although the installation of automatic toll machines was considered in the fall of 1955; toll collectors claim that an experienced collector can conduct the operation faster than a machine.

In at least one case, striving for speed in toll collection fouled up the operation, as Hiram Young, Supervisor of Maintenance, pointed out. The Authority sold truck commuter tickets to companies which

Interstate and Memorial Bridges. Portsmouth, N.H., in foreground; Kittery, Maine, in background. Notice the original black paint of the Memorial Bridge (now bronze green like the Interstate). The primary reason for the new bridge, downtown Portsmouth congestion, becomes perfectly clear in this picture. October 9, 1947. Photographer facing west. Photograph by Paul S. Otis of N.H. Dept. of Public Works and Highways, courtesy of Roger Staples.

used the bridge frequently; the trucks would go through on the outside lane at 50 m.p.h., and the collector would pull the appropriate ticket and punch it. The truck operator would then be sent a monthly statement. Things would get so hectic that collectors often missed seeing the truck, so they would make an educated guess and pull a ticket. One month, they sent a statement for a number of trips to St. Johnsbury, a Vermont company, which replied that it hadn't had any trucks go over the bridge that month. Things tightened up after that.

U.S. Route 1 Bypass; Portsmouth, N.H. Overpass carrying Woodbury Avenue. 1950 dedication plaque on the north side of the handrail pier, west end, south side of Woodbury Ave. George R. Laderbush was a Portsmouth native, killed in the South Pacific in World War II when the submarine Flyer struck a mine. Photograph by author.

Otherwise, toll collectors looked on as traffic piled up during summer weekend lifts, sometimes as far south as Hampton, New Hampshire, and endured the usual dumb tricks pulled on toll collectors as well as people riding through in various states of disrepair, so to speak. Today, the toll collection is memorialized solely by the wide spot in the bypass pavement south of the bridge where the toll booths stood. As soon as the toll booths disappeared in 1972, this apron became a popular, unofficial parking place for patrons of the nearby Hanscom's Truckstop (the running battle between Buzzy Hanscom's empire, located on both sides of the Bypass, and the Bridge Authority, appears as a continuing theme in Authority minutes, which periodically note the mysterious reappearance of a hole in the median fence).

Shortly after World War II, a number of changes were made in the bypass. A sore spot with Portsmouth drivers was the grade-level crossing at Woodbury Avenue, which was hazardous enough to require a police officer directing traffic on weekends. One of the promoters of an overpass at this location was William J. Linchey, who later became the Portsmouth City Marshal. Another was David Smith, Street Commissioner of Portsmouth, who owned land along Woodbury Avenue; he offered to sell to the Authority whatever additional land was needed. At a meeting of the Bridge Authority on September 6, 1949, approval was given for connecting the New Hampshire Turnpike (now part of Interstate 95) with the Route 1 Bypass, and for constructing an overpass to carry Woodbury Avenue in Portsmouth over the Bypass, thereby eliminating the intersection (i.e., grade crossing) of the two roads at that point. The overpass was constructed in 1950 and named in honor of George R. Laderbush, a native of Portsmouth who was killed in World War II while serving aboard the submarine *U.S.S. Flyer* in the Mariana Islands. Laderbush, a graduate of Portsmouth High School, had been an electrician apprentice at the Portsmouth Naval Shipyard before his enlistment in the Navy. His name appears on a plaque facing Woodbury Avenue on the south side of the overpass at its west end.

The creation of the Portsmouth Traffic Circle as part of the project necessitated the moving of Howard Johnson's Restaurant from the northwest quadrant of the circle to its present location on the east side, because it was in the way of the projected ramp from the southbound bypass to Woodbury Avenue. The building was moved by the W.B. Hill Company of Tilton, New Hampshire, on December 29, 1949. The Portsmouth traffic circle was constructed at the same time, as was a spur road paralleling Woodbury Avenue from the circle and leading to the General Sullivan Bridge in Newington, New Hampshire. This coincided with the construction of I-95, which was built in 1950 and came into the new circle. The spur road would become part of the Spaulding Turnpike when the latter was built in 1956.[37]

At this time, Dr. Paul's dog cemetery disappeared, covered up by the northbound exit from the bypass to Woodbury Avenue. On the Kittery side, I-95 joined the bypass just north of Remick's Corners, as it does today, and the movement was afoot by 1951 to reorganize the bridge access. An Authority meeting of December 5, 1951, noted a proposal to relieve Navy Yard traffic congestion by constructing a ramp from the intersection of Bridge Street with the Bypass, southerly along the east side of the bridge, and looping under the bridge to provide access to the southbound lane of the bridge approach. This "would necessitate the rebuilding of the old wooden overhead railroad bridge at the intersection of Bridge Street and the Old Post Road..." It was further noted that the Commission "has already given consideration to an access road extending from the Navy Yard Street somewhat parallel to Government Street in Kittery, and connecting with Route 1 and an extension of this route as a secondary project to connect with

Interstate Bridge with northbound Boston and Maine train, locomotive #3702, at the Kittery end of the bridge. C.1952. Photographer facing south. Ray Tobey Collection.

State Route 103." This project was carried out, called by the name of Oak Terrace, a street which was incorporated into the loop. The project resulted in the demolition of the Holt and Hurd houses on the west side of the bypass near the bridge; these had survived the original construction, although they were very close to the bridge and approach. The Oak Terrace project removed much of the hillock upon which they sat, right down to the ledge. The project was intended to prevent the dangerous left-turn of traffic from Bridge Street in Kittery onto the bridge heading south, particularly by trucks, but has not entirely stopped this practice.[38]

On November 1, 1949, a ceremony marked the end of tolls on the General Sullivan Bridge at nearby Dover Point in Newington, New Hampshire, with John E. Holden (Newington oil dealer who, apparently, had "bought the bridge" with his toll payments over the years) paying the last fifteen cents to Governor Sherman Adams of New Hampshire. The *Portsmouth Herald* of that day also reported that Frank E. Brooks of Portsmouth, Treasurer of the Interstate Bridge Authority, had suggested that Interstate Bridge tolls might be lifted in fourteen years, if traffic continued to increase at a rate of 2 to

Interstate Bridge. The look of speed: Boston and Maine train #250 headed south on May 15, 1952. The last year of operation of the Eastern Division of the B. & M. Railroad through Portsmouth and Kittery. Ray Tobey Collection.

3 percent annually, so that the bonds could be retired by 1963. He noted that traffic had increased 13 percent over the 1948-49 fiscal year ending June 30.

As early as 1947, the Authority became very concerned with the sudden increase in requests for access to the bypass. The board consulted its legal advisors, Varney & Fuller, whose opinion, dated January 26, 1948 (and attached to the I.B.A. Minutes of the January 16, 1948 meeting) referred to the New Hampshire Supreme Court ruling in the case of the Interstate Bridge Authority v. Ham Estate, 92 N.H. 277. In this case, the Court noted that the Authority is a public corporation, "of which the two states are the sole members, and they are the real owners in trust for public use." It went on to say that the bridge approach, outside of the toll booths, is a public highway and that abuttors "have full right of access thereto, except within the toll-gate limits, where they are subject to rules and regulations governing the use of the bridge and any other services made available in connection with said bridge." As a result, at this meet-

The Flying Yankee *train in the Portsmouth, N.H., station. It went from Bangor to Boston on the Eastern Division of the Boston and Maine Railroad. Its streamlined design is characteristic of the era when the Interstate Bridge was built. Bob Shaw photograph.*

ing, the Socony-Vacuum Oil Company was granted access to the highway at a point on the southerly side of the New Hampshire Approach to the bridge. Many other requests for access—for gas stations, tourist cabins, a refreshment stand and a lumber company—were granted in 1949 and 1950.

Maintenance costs began to rise in the early 1950s, due to rapidly increasing traffic, with a projected deficit of $4,774.41 in the 1951-1952 budget. Before long, the tolls caught up with the maintenance costs. The fact that The Authority advertised in *The Trail,* the official magazine of the Maine Good Roads Association, and *New Hampshire Highways,* the official magazine of the New Hampshire Good Roads Association, was more of a goodwill gesture than an effort to promote use of the bridge. After approving salary increases of ten percent for Maintenance and Bridge Operation Personnel, and purchasing a sandpit and payloader, the Authority asked Attorney George D. Varney to take steps required for issuing additional revenue bonds. This was done as part of the refinancing of the bridge in 1956.

The fifties were a period of change in the transportation sys-

Refinancing of the Interstate Bridge. Closing ceremonies, with Andrew Jarvis, Clerk of the Maine-New Hampshire Interstate Bridge Authority, receiving the check for $2,103,797.50 from J. C. L. Tripp, President of Tripp & Co., Inc., underwriters of the bond issue. Seated to the left of Mr. Jarvis is George D. Varney, Attorney for the Authority. Standing (left to right): Harry W. Faath, Vice President of Tripp & Co.; W.S. Philibert and Emil J. Horn of Chemical Corn Exchange Bank, Paying Agent; Richard Petty of Mitchell, Pershing, Shetterly & Mitchell, Revenue Bond Attorneys; and Walter N. Burnett, Assistant Cashier, The First National Bank of Boston, Trustee. On their way home, Messrs. Varney and Jarvis stopped at the latter's favorite Boston restaurant. Mr. Jarvis asked the owner if he could cash a check and received a reply in the affirmative. The gentleman was surprised when he looked at the check. Photograph courtesy of Ms. Catherine P. Jarvis.

tems of Maine and New Hampshire. The last scheduled train (a passenger train) of the Eastern Division of the Boston and Maine Railroad left Boston for Portland on September 27, 1952. This last eastbound train was #21, with Engineer Mahlon Phillips, Conductor Joseph R. Boyce. The last westbound train, earlier in the day, was #250, with Engineer G.W. Caldwell. After that, some unscheduled freight trains may have used the track between Kittery and South Berwick, Maine; by the time of the 110th anniversary of the Portland, Saco and Portsmouth Railroad on November 25, track removal was largely completed.[39]

In 1956, major improvements to the bridge approaches were carried out by Landers and Griffin, Inc., of Portsmouth, and Morrison-Knudsen Co., Inc., of New York City, operating as a joint venture; the amount of the bid was $548,243.55. New toll booths were added, the approaches were repaved, and new entrances and exits were constructed. Increasing traffic had created the need for another traffic lane and traffic lights; because the heaviest traffic was (and is) northbound on Friday nights and southbound on Sunday nights, the middle lane was made reversible by the toll attendants. Although each of the three lanes was only 10' wide, accidents were few and mostly minor. After the opening of the I-95 bridge in 1972 reduced traffic on the Interstate Bridge, the middle lane became unnecessary; today, the middle lights are inactive and await removal. To pay for these improvements, the bridge was refinanced with $2,100,000 in Bridge Revenue Refunding and Improvements Bonds, underwritten by Tripp & Co. of New York.[40]

As revealed by the Tripp Prospectus, traffic over the bridge had increased rapidly between 1946 (1,511,263 vehicles) and 1954 (3,230,383 vehicles). During this period, net income increased 135 percent—from $113,672 to $267,051. Included in this income was an annual rental payment of $41,000 from the Boston & Maine Railroad. Notable is the fact that operating and maintenance costs had increased only 41 percent during this time. Thus, the financial picture for the Authority had improved, particularly since the early 1950s—a good thing, because light wartime traffic had meant that the tolls had barely covered the bond payments.

Along with highway traffic, marine traffic was increasing rapidly during this period. This led to many more lifts than were envisioned when the bridge was opened and, as a result, increasing concern on the part of the Authority with highway traffic delays and wear on the bridge itself.

These concerns were expressed to Harrington & Cortelyou during their annual inspection of the bridge in 1963 and they proposed installing a movable small craft span in the railroad deck. At the meeting of April 1, 1963, the Authority authorized the chairman to contact E.M. Newman of Harrington and Cortelyou to obtain from him a proposal to study the feasibility of cutting or altering the railroad tracks to improve small craft navigation in the river.

The engineers recommended that the southernmost steel girder span at the Kittery end of the bridge be made movable so that it could slide into the adjacent truss span, thereby creating a small boat passage 70 feet wide between fenders and 8 feet deep at mean low water, with a vertical clearance of 36 feet at mean high water or 44 feet at mean low water. Thus, small boats could pass through the bridge without opening the main lift span and interrupting highway traffic. Comments on the proposal were invited by the U.S. Army Corps of Engineers on May 4, 1964. The $400,000 cost was to be paid from reserve funds of the Authority; of course, the change had to be approved by the Maine and New Hampshire legislatures and the U.S. Congress.

Built in 1967-1968, the movable railroad span rests on two platforms, one at each end, each having two nuts on either side, as well as a short length of railroad track. Thus, the span operates with eight large screws, two at each corner, each one surmounted by a nut about 12" in diameter and an electric motor and gear box. The south end rests on standard railroad wheels, with the north end suspended by arms. The arms rest on wheels set on the northern pair of short lengths of railroad track, the track being slightly inclined southward; above each of the two arms is a hole cut in a trolley beam. The operator pushes a button activating the motors, the span travels up on the screws, and the arms (with wheels at the tops) go up through the holes cut in the trolley beams. When the span reaches the top, the operator pushes another button activating the drive motor, which is connected by a chain drive to sprockets on the railroad wheels at the south end of the span. These pull the span open. The operation is reversed to close the span. The screws and nuts operate like large jacks—say, house jacks, with the screws staying in place while the nuts ride up or down on them. Once the span is raised, it is supported by the railroad wheels on one end and by the arms on the other, with an internal motor to move it into the adjacent span. The control panel is located on a platform midway between the two decks.[41]

Several 'bugs' were discovered in this unique movable span after it went into operation. The most serious problem was the way in which the nuts were lubricated. As originally designed, the steel screws and bronze nuts were greased. What could not be foreseen was the fact that their environment would be a very dusty one; added to the usual road dirt was a substantial amount of dust from the nearby gypsum plant. Collectively, this built up in the grease and turned it into a grinding compound; one day when the span was

Interstate Bridge. Movable small craft span in the railroad deck raised and rolling into the next span southward, with Piscataqua River Bridge under construction behind. C.1970. Photographer facing west. Courtesy of Harrington and Cortelyou.

raised, a nut fell out, having disintegrated. The engineers collected four or five large jacks, and let the span down a little at a time, alternating jacking with using the other seven screws. Resulting modifications included refinishing the screws with a finish considered more suitable, offsetting the nuts, realigning the span, and arranging to have the nuts operate in an oil bath rather than grease. Notwithstanding these initial problems, encountered in an admittedly highly unusual piece of equipment, the modification of the girder span has saved the Bridge Authority thousands of dollars in maintenance costs to the main span equipment, not to mention the costs of fuel and aggravation which would otherwise have fallen on drivers stopped for bridge lifts.[42]

Other modifications in the 1950s included repairing the bridge and approaches. They were originally paved with concrete, with the lift span incorporating a lightweight compound called Haydite, which used pieces of baked clay instead of rock for the aggregate. Bridge and approaches were repaved with asphalt in the early 1950s. By the mid-1960s, the underlying concrete had started to disintegrate. To repair 2,880 feet of road, the Bridge Authority spent $228,930 in 1965-66. The contractor was Norman E. Jackson, Inc. of Pittsfield,

Interstate Bridge. Repaving of bridge. Spring 1987. Photographer facing south. Photograph courtesy of Public Information Office; N.H. Department of Transportation.

Maine. The concrete deck was patched and again overlaid with asphalt. Also in December 1965, the Bridge Authority approved traffic lights at Cottage Street and Greenleaf Avenue in Portsmouth, but turned down a request, supported by City Manager Robert C. Violette and State Rep. William F. Keefe, to remove the median strip of the bypass from Coakley Road to a point "150 feet from the railroad overpass" so that businesses could have better access. Engineers advised the Authority that removal of the strip would contribute to hazardous conditions on the bypass.[43]

The bridge was redecked again in the spring of 1987, this time patching the underlying concrete and bringing it right up to the surface, as originally paved, instead of using an asphalt overlay; the contractor is Atlantic-Corey Inc. Another lightweight concrete was used for the lift span. Ernest Searles reports that the 1966 concrete remains sound.

In 1968, the ropes connecting the counterweights and lift span were examined for wear by a wire rope specialist, John A. Herr of Branford, Connecticut. He noted that the original ropes were still in service (as they are today) and observed, "The lift span is supported by 64 ropes - 16 on each corner - 1 3/4" diameter 6 x 19 filler wire (6x25) construction plow grade, preformed with fiber core. Each rope is 160 feet 7 1/4" long out to out of bearing sockets, length measured under a load of 9,200 pounds." The ropes go over bull drums (four in all) 10' in diameter, which each have 16 grooves, and serve to deflect the ropes from the sides of the span to the counterweights as well as drive the lift operation. A sharp report was heard during a lift, which Herr attributed to the ropes not being equally tensioned, so that some ropes slid on the bull wheel in an effort to catch up with others; presumably the report was heard when the laggards finally caught. This was said to possibly cause wear on the bull wheel grooves, which would thereby end up being of uneven diameter or circumferance. Herr recommended that the grooves be reground to have the same dimensions when the ropes were replaced. The Herr report was an early indication of problems with vibration and the span going out of skew during a lift; around this time Selsyn motors were installed in the towers to measure the height and level of the lift span by their connection with the bull drums and transmit this information to the operator; by this means, he can determine which end of the span is up or down and adjust it so that it will seat properly when lowered. Specifications and drawings by Harrington and Cortelyou for shims on the counterweight and lift span ends of the lift cables, as well as a report by David Dickey at the Bridge Authority's March 1977 meeting, show that the Authority was still trying to solve the vibration problem. At this time (1987), the cables have been shimmed as much as possible and are showing definite signs of wear; they are due to be replaced, for the first time, in the near future. Meanwhile, vibration has continued to be a problem for the bridge, and the effort to minimize its effects and keep the span level accounts for the slowness of lifts—a situation well known to motorists.[44]

At the meeting of December 18, 1970, the Bridge Authority decided to rebuild the operators' house on the bridge, on the recommendation of A.R. Sirois and Edward T. Swierz, bridge engineers for Maine and New Hampshire respectively. Harrington & Cortelyou, consulting engineers, and Bernard H. Langley suggested the use of stainless steel for movable parts and Cor-Ten A steel for the siding. A second story was added to the house, in order to provide some living quarters for the operator and gatemen (whose shifts are normally 12-hours, except on Wednesdays) and better lines of sight for highway and river traffic. Another improvement carried out in this period was the installation of a revised Selsyn Transmitting and Indicating System for determining whether the bridge is out-of-level, installed in 1972.[45]

Despite constant maintenance, the bridge has been temporarily disabled many times over the years. A photograph published in the *Portsmouth Herald* (April 22, 1971) shows a Public Service truck backing out on the railroad deck to replace a failed transformer which prevented the lift from operating. The failure occurred after a power outage in Portsmouth, scheduled so that the company could change lines in the city to increase voltage.

That spring, on May 24, a Corps of Engineers hopper dredge hit the bridge. The 351' *Comber* had been dredging the channel off the Simplex plant and was heading downstream for a dumping area just beyond the Isles of Shoals when it was caught in the ebb tide and hit one of the bridge abutments at about 6:05 p.m. The bridge operator saw it heading straight for the operators' house and ran; as it turned out, the building was not hit. Damage to the bridge was minor (spalled concrete and some lead wool joint filler pulled out as the vessel scraped along the face of the pier), but the dredge dumped its load off the state pier in about ten minutes and sealed its bulkheads "to prevent possible sinking."

The *Comber* accident was one of a number of bridge hits by barges, most of them minor. One more serious accident occurred when an oil barge hit the bridge, stuck for a whole tide and spilled a lot of oil. From then on, barges were required to carry pilots when navigating the Piscataqua River. Hiram Young, Maintenance Supervisor, was riding up on one of the counterweights when an oil barge struck—a frightening experience indeed! He said the barge scraped along a piece of steel sticking out from one of the cofferdams (used to build the bridge and left in place as part of the piers); this steel edge opened it up like a can opener. Kevin Guy, Portsmouth realtor, recalls another hit by an oil barge late one night; he was heading north and had to stop on the bridge for the lift. It was a nice night and his was the first car, so he got out and stood by the barricade to watch the ship go through. It hit and "the whole bridge shook; the barge tore a hole in its side about 70' long and spilled a lot of oil, which went up and down the river." The oil coated boats and everything else in the river; it was the first real oil spill and occurred around 1963 or '64.[46]

On at least two occasions, bridge malfunctions caused traffic jams. One occurred on June 9, 1971, when the lift span on the bridge stuck in the raised position during morning rush hour traffic. Following this, Harrington and Cortelyou, consulting engineers, inspected the lift span on June 16, 17 and 18, but did not find anything in the electrical systems and equipment that could account for the malfunction. Although they noted minor damage to the cross bracing of the south tower, they subsequently concluded that it was due to the span having gotten out of level at some time in the past, rather than to the

Comber collision.[47] Until 1972, the Bridge Authority hired Harrington and Cortelyou, consulting engineers, to carry out annual inspections, which the firm did that year. Beginning in 1973, inspections of the bridge and approaches were done annually, between April 15 and May 15, by engineers from the two state highway departments. The consulting engineers were called in for large projects and special problems. It has been said by some that maintenance slipped after this, but this claim would be hard to substantiate; more plausibly, as the bridge got older, it required more attention.

By 1974, the lift span gear reducer units (one in each tower) were in need of replacement. The Bridge Authority consulted Harrington and Cortelyou, and Frank M. Cortelyou, Jr. summarized findings made during a visit to the site on June 17. Earle Gear & Machine Company had made the original units, and (in a letter of October 21, 1971) had suggested replacement of one and reconditioning of the other. Mr. Cortelyou said there was evidence that "the two end bearings for some shafts had their sleeve bushings bored not on a horizontal axis." He also compared reducers of the original type with modern ones, saying "As near as we can determine, all standard present-day reducers as carried in manufacturers' catalogs for the type and size here involved are made with cylindrical roller bearings rather than with the bronze sleeve liner or bushing type in the present units. Also, improvements over the 35 year age span of these units in materials for gears and shafts, and in tooth cutting, have resulted in smaller units of less weight for meeting the same performance criteria." Ultimately, the contract was awarded to the D.O. James Company of Chicago for two new units, which were installed in April 1976 in an eighteen-day project—the first time the bridge had been out of service for that much time. A panel was cut out of each machinery house so the reducer could be removed, and was replaced after installation of the new one.[48]

The vibration problem, mentioned in Authority records as early as 1968, came to the fore again in 1976. Dr. Eric E. Ungar, Principal Engineer of Bolt, Beranek and Newman, Inc. (Cambridge, Ma.), studied the problem in August of that year; his observations are interesting because they illustrate the particular problems associated with a vertical lift bridge. The following is taken from his letter to David V. Dickey, Assistant Adminstrator, Public Works Division, State of New Hampshire (August 27, 1976):

> The vibrations I observed in the machinery house of the south tower consist of motions essentially in the upstream-downstream direction at an estimated frequency of 3 cycles per second, with estimated amplitudes of up to 6 inches. These

vibrations occurred primarily during the first few feet of lift. No significant vibrations were set off during lowering of the lift-span or in test runs where the span was accelerated upward from rest at about mid-height as much as the machinery would allow.

It takes a force, moment or impulse that essentially acts in the upstream-downstream direction to move the tower in that direction. Significant causes of such motions in the bridge are difficult to visualize. However, the effects of relatively small excitations are amplified by structural resonances, if these excitations occur at the proper frequencies. As you know, reduced damping increases the amplification at resonances; thus, your improved lubrication program, which undoubtedly has led to a reduction in the system friction, may have led to an increase in the vibrations.

My best present hypothesis is that the vibrations are caused by the lift-span being picked up unevenly - e.g., first on the downstream and then on the upstream side - causing the tower to deflect to one side (e.g., dowstream) and then to oscillate. As the tower oscillates in this manner, one would expect to obtain a simultaneous vibration of the lift span, involving oscillatory rotational motion about its longitudinal axis. My calculations indicate that this motion occurs at about 3 cycles per second, which matches the observed frequency. As I observed the cables on the south tower, I noted that the 16 downstream cables all oscillated together when they picked up the load, but that some of the 16 upstream cables were taut while others flapped loosely. This leads me to suspect that only a part of the upstream bundle of cables carries the load initially, making this bundle in effect less stiff than the downstream bundle, and leading (at least during the initial part of the lift) to an unequal upstream-downstream distribution of the load. The popping sounds that come from the upstream cable may be another indication of this uneven initial distribution of loads among the 16 upstream cables.

The 1977 Annual Inspection Report on the bridge, compiled by the Maine and New Hampshire bridge engineers and maintenance supervisors and Bridge Authority personnel, reported that re-balancing the counterweights and adjusting the cables had produced a noticeable decrease in vibration, although occasional re-levelling of the span was still necessary. This meant that normal operating speed could be resumed.

Electrical power problems with the lift span were reported in the spring of 1982, and David Dickey of the New Hampshire Department of Public Works and Highways arranged to have the power cables tested. Very poor performance was discovered, and it was found that older submarine power cables, disconnected some years earlier, would give adequate service. Likewise, out-of-level problems with the span were discussed more than once; it is safe to say that both the power transmission and levelling problems were related to the aging electrical system of the bridge, which has never been completely replaced. The specifications for complete replacement of the system are being drawn up by the Authority's consulting engineers, and it is expected that the work will be completed in the next several years. As Harrington and Cortelyou have stated, there is nothing to be gained by partial replacement of the system.[49] One detects a certain amount of anticipation as the operators talk about the computerized system they have heard will be installed.

The electrical problems were highlighted by an incident which occurred during a September 1987 raising of the lift span to its full height. As the operator explained it, he raised the lift as usual, pushing the bypass button to operate it as it neared the top slowly. To his surprise, the south counterweight kept coming down—all 750 tons of it—and bounced off the deck. He called the gateman on the north side, who reported that his counterweight was above the door in the tower (about 10' up), where it stopped when the brakes were applied. An electrical relay had broken, so that the south brake did not work, and the counterweight kept coming down. Clear evidence that the electrical system needs to be replaced! Needless to say, the environment of the Interstate Bridge is a very tough one for an electrical system—salt water, road salt, water and moisture (including fogs), and extremes of weather.

A problem with the lift span in January 1987 was due to the particular problems associated with operating a bridge of this type in a northern climate. The logbook tells the story. At 6:38 p.m., the lift was raised fifty feet for two fishing boats, which passed through at 6:43 p.m. The tide was flooding and it was snowing at the time, with a temperature of 41°. Under Remarks was written, "Bridge will not seat. Goes out of skew." It was finally seated at 8:30 the next morning. The wind was blowing hard from the northeast at the time the span stuck (described locally as "a Northeaster"), with freezing conditions, forcing it against the guides on one side, where it evidently froze instantly. By the next day, the wind had shifted to the northwest, with the weather described as clear, and the operator was able to lower the span. The electricians tested the span with several lifts, but found nothing wrong.[50] A more common problem is to get ice and snow on the contact fingers where the lift span rests on the deck; this can prevent electrical contact, so that the green lights won't operate and the gates won't go up; in other words, the roadway and railroad cannot operate until the problem is solved. Because the contact fingers are down on the railroad deck, the railroad is supposed to have maintenance men on the scene, but has never done so, according to a retired operator.[51]

Just completed by the Bridge Authority is a $1,347,045.00 project carried out by Atlantic-Corey and Subsidiary, which has included a concrete redecking (the original deck was concrete; the last one was asphalt), repairing the concrete piers, and painting where needed.

VI *Tolls End on the Dime Bridge*

A New Era for the Authority

BY 1970, RESIDENTS were looking forward to the opening of the new Piscataqua River Bridge, which would close the gap in high speed travel on I-95. A *Portsmouth Herald* article of August 28, 1970, remarked that the "ghastly figure" of 43,350 vehicles had passed through the tolls the previous Sunday, "described as a 'typical day' on the bridge by a worker at the Maine-New Hampshire Interstate Bridge Authority office at the bridge toll booth..."

The Piscataqua River Bridge was opened on November 1, 1972, with David H. Stevens, Commissioner of the Maine Department of Transportation, organizing the ceremony and cutting the ribbon. Although the new bridge was welcomed by most people, merchants along the bypass feared a drastic decline in sales of gasoline and cigarettes would result from the bypass being much harder to find.

In fact, some merchants were hard hit. When the "high level bridge" opened, gasoline sales at one station (of thirteen) dropped from 1200 gallons per day to 500, and cigarette sales from 200 cartons to less than 90. This represented a loss to the State of New Hampshire of seven cents per gallon of gasoline and $1.10 per carton of cigarettes.

For the Interstate Bridge itself, the opening of the new bridge would have the immediate effect of decreasing traffic to the point where toll collection would be unprofitable. For this reason, tolls were ended at midnight on October 31, 1972. By November 6, the toll booths were being demolished. The opening of the new bridge and the removal of the tolls marked the end of an era for the Bridge Authority and, in a sense, the beginning of financial crisis for that body, although the outlines of the trouble would take some years to become clear. In the words of Wallace F. Stickney, New Hampshire Commissioner of Transportation and Vice-Chairman of the Authority, "When they stopped collecting tolls, the die was cast. The point will be reached when the Authority can't generate enough money from the trust funds..." Today, the visible effects of the new bridge on the bypass are evident. Businesses recovered from the change, grew and flourished, while the road itself became more of a service stop for trucks and cars using I-95. The elimination of tolls also encouraged the use of the Interstate Bridge by local traffic. Thus, the whole character of the bypass has evolved from that of an express highway to a truckstop and connector of local roads. This parallels the transformation of U.S. Route 1 the length of the East Coast in response to similar circumstances.

Several themes run through the history of the Authority in the 1970s and 1980s. Among the most prominent are the increasingly bleak financial outlook for the bridge beginning about 1980, and the virtual disappearance of the Boston and Maine Railroad, which were to a small degree related.

The railroad had paid an annual rental fee of $41,000 beginning with the opening of the bridge in 1940. Having discontinued its Eastern Division line through Kittery, Maine, in 1952, it had renegotiated the rent in 1959. In 1970, the railroad went bankrupt and stopped making payments. The issue was raised at the I.B.A. meeting of November 4, 1981, when Attorney Hancock was instructed to explore the possibilities of collecting the unpaid rent and future rental payments from either the railroad or the Navy Department (the Portsmouth Naval Shipyard being the only user of the railroad link since 1952). At the meeting of April 22, 1983, it was revealed that a settlement of $165,931 for railroad rent through July 1, 1982, was expected; the Authority had, theoretically, been receiving about $12,000 per year in rent. The City of Cambridge, Massachusetts, apparently also in line for payment, had withdrawn its objection to the settlement between the railroad and the Authority. Guilford Transportation's purchase of the Boston and Maine in 1983 improved the financial outlook for the railroad; subsequently, the Bridge Authority passed a resolution reducing the railroad rent to $7,000 per year, with the possibility of an increase if railroad traffic increased within five years. The fee was eventually set at $7,500 per year, beginning in 1983. However, at the meeting of September 7, 1984, Roger Staples Treasurer, reported that the Authority had not yet received the rent for the previous fiscal year.

Not only was the Authority affected by the general unprofitability of railroads in mid-twentieth century America, it also suffered the consequences of a highly litigious social environment when it tried to secure insurance coverage. To meet the rising cost of insurance, the Authority had to raise the 1984-85 operating budget $20,000, because the insurance premium jumped from $40,250 to $60,250. Blake Insurance Agency meanwhile reported that it was having great difficulty finding insurance carriers who would provide both general liability and umbrella coverage. At its August 30, 1985 meeting, the Authority approved payment of $25,923 for a Comprehensive General Liability Policy to run to September 1, 1986. However, it still sought to obtain an Umbrella Liability Policy. This kind of policy was finally obtained, but was cancelled by the carrier on January 17, 1986, after only three months. Thus, the insurance issue became a very serious problem for the Authority in this period. This compares with the insurance situation when the bridge was being built; the Authority requested information from the Maine Highway Department regarding insurance for toll bridges and discovered that the ones in that state were insured only for the loss of toll money![52] As noted elsewhere, war risk insurance carried by the Authority is discussed in a letter dated April 6, 1944, from Mr. Purrington, Executive Secretary of the Authority, to Mr. Wilder of the Maine Highway Department.

Two large capital improvement projects approved in 1980 characterized the year in which the budget crunch became a reality. On March 21, the Authority accepted the bid of W. G. Wheeler, Inc., for sidewalk reconstruction on the bridge for $209,899.50, and on June 21 of that year, Maynard L. Young's bid for painting the bridge for $208,000.00 (the total cost, including paint, would come to approximately $282,000). The significance of these large maintenance bills will be discussed below, in relation to the reorganization of the Authority in the 1980s, to which they served as a prelude. It is probably safe to say that this was the most difficult issue confronted by the Authority since the original financing and construction of the bridge; no other problem has taken up so much time or been so difficult and time-consuming to resolve. First, however, it is important to discuss the maritime traffic through the bridge.

VII *Traffic Through the Bridge*

ALTHOUGH THE BRIDGE provided a valuable service to vehicular and rail traffic, from the point of view of tugboat captain Richard Holt, it constitutes a real problem for river traffic: "The Interstate Bridge is a particular hazard because it crosses the channel at a bad angle...The old bridge had an opening only fifty feet wide, so the 200' width of the new one sounded big when it was built. The Interstate Bridge limits the size of ships by its vertical and horizontal clearance; it is 200' wide 'straight,' but it sits at an angle of thirty to forty degrees, so you probably lose forty or fifty (feet)." He added, that "a dangerous feature of the IBA Bridge is the piers." They have a step about fifteen feet underwater, which could cut a hole in a vessel's hull underwater, so that's "something they always have in mind when going through the IBA Bridge." Because of the current, pilots endeavor to schedule the passage of ships through the bridges at slack tide. To make matters more complicated, high tide occurs two hours earlier in Portsmouth Harbor than upstream in Great Bay, so that the water level around the two bridges can already be dropping while the water continues to flow in, filling up the bay.[53] As the ships passing through the river have increased in size, so too have the navigation problems. In the 1950s, the standard size for ships coming into Portsmouth Harbor was less than 20,000 tons; now there are very few less than that and the norm is 50,000 to 55,000 tons. Many of these ships are oil tankers. When the harbor first began to get the larger ships in the mid-1970s, a committee was formed to study navigational improvements (the Portsmouth Harbor Safety and Improvement Committee, formed in 1976—Capt. Holt, Chairman), its object being to make the harbor safer for bigger ships. Federal and state

Tug Eugenia Moran towing the Gold Bond Trader upriver on October 23, 1987; deckhand visible to right is Don Howcroft. Pilot on the ship is Captain Dick Holt (not shown). Piscataqua River Bridge behind the Interstate Bridge. Photographer facing west.

funding was approved for the project, which includes widening the channel on either side of the Interstate Bridge. A related project, still in the planning stage, is the building of a docking facility west of the Interstate Bridge, between the bridge and the gypsum plant. The new facility would be in a prime location, because it would have access to river and rail transportation as well as I-95. Of course, this would tend to increase traffic through the Interstate Bridge.

Many factors have limited commerce on the river; for example, dock space is at a premium. Docks are so expensive and hard to come by, that very few single-user docks remain; most have three or four customers now. The height of the bridges has also discouraged commerce. Although the 135' height of the Interstate Bridge was pretty standard in 1940, since that time, the increase in the size of ships has made this height problematic. Nevertheless, the I-95 bridge, located a short distance upstream from the Interstate Bridge, was opened in 1972 with a 135' clearance, based on the reasoning that it made no sense to provide more clearance upstream from the former bridge. The navigation committee considered all of these factors, but particularly the width limitation of the Interstate Bridge, which could be alleviat-

The Gold Bond Trailblazer *has passed through the bridge, heading for the National Gypsum Co. plant. South tower of Memorial Bridge in background. Note angle of the channel in relation to the Interstate Bridge. October 23, 1987. Photographer facing east. Photograph by author.*

ed. The committee looked at the possibility of rebuilding or replacing the Interstate Bridge to lengthen the lift span, but decided that reconstruction was impractical. The Corps of Engineers considered the issue in the spring of 1984 as part of the dredging project, and concurred in the committee's decision.[54]

The process by which a ship passes through the bridges is relatively complex. First, the ship contacts the bridge operators by VHF radio (ships began using VHF radios in the 1960s; Channel 13). The tugs may call on the land telephone if they are going to go through early in the morning, or to check-in when conditions are bad and the bridge might have trouble functioning. Likewise, the bridge operators contact the Portsmouth Navigation Company, the tugboat operators, if weather conditions or other factors may prevent opening of the bridges. The pilots are particularly concerned about poor visibility, usually caused by fog, and very high winds. The area encompassed by the Interstate and Memorial Bridges is the most dangerous part of the river, with the passage through the Interstate Bridge the worst spot. Therefore, good visibility is considered essential for passage. The tugs communicate with the pilot on the ship by radio and acknowledge his

Interstate Bridge. James Howard, operator, raising the lift span. September 4, 1987. Photographer facing north. Photograph by author.

commands by one whistle. Other whistles are given to push (one), to signal that one is backing (two short), and to signal that one is casting off (three—one long and two short). The tug may cast off from one side of a ship going through the Interstate Bridge if the pilot considers the ship too wide for passage with tugs on either side.[55]

Procedures for opening the Interstate Bridge are governed by the River and Harbor Act of August 18, 1894, and are published in the *United States Coast Pilot.* The paragraph on the Interstate Bridge gives the signals for its opening, which are two long and two short blasts of a whistle, horn or megaphone. When the draw can be opened immediately, the draw tender replies with "two long distinct blasts;" when it cannot, or if it is open and is to be closed immediately, the draw tender replies with "five short distinct blasts of a whistle, siren, horn, or megaphone or by five loud and distinct strokes of a bell, repeated at intervals until answered in like manner from the approaching vessel." When weather conditions prevent these sound signals from being heard, "signals shall be made from the vessel by swinging in circles at arm's length, a lighted lantern at night and a flag by day," and the draw tender shall reply by raising and lowering a lighted lantern or flag a number of times (if the draw can be opened immediately), or by "swinging to and fro horizontally a number of times" a lighted lantern or flag (if the draw cannot be opened or is to be closed immediately). Similar signals pertain to the Memorial Bridge, with the signal for

Interstate Bridge. Operators' house. Lift span height and skew indicators. November 1987. Photographer facing south with his back to the railroad panel. Photograph by author.

opening being "four blasts of a whistle or horn." In practice, most communication between ships and the bridge is carried out over the radio.

After receiving a signal from a vessel, the bridge operator follows a set sequence for raising the lift span. After judging the highway traffic situation manageable, he activates the control board. This is done by flipping the Gate Control Bypass. Then he pushes the buttons which control the land traffic lights on the north and south approaches to the bridge and put the land gates down, stopping traffic from going onto the bridge. The operator then sounds a bell on the railroad deck, warning anyone there that a lift is about to begin, and puts the span gates (the gates near the lift span) down. Finally, he sounds the siren, which is the final warning before the lift takes place. Grasping the handle, sometimes called a tiller or throttle, in the middle of the panel, he begins the lift. Due to the age of the mechanical and electrical systems, the lift is executed as slowly as possible so that the span will not get out of skew. The appropriate entries are made in the log, and after the vessel passes through, the operator looks to make sure nothing else is approaching or has sneaked into the draw; then he sounds the siren just before the lift comes down. At approxi-

Interstate Bridge. Operators' house, first story. Lift span control panel. November 1987. Photographer facing east. Photograph by author.

Interstate Bridge. Operators' house, first story. Railroad circuit board to left, reflective vest hanging on it. Cot at bottom. November 1987. Photographer facing north. Photograph by author.

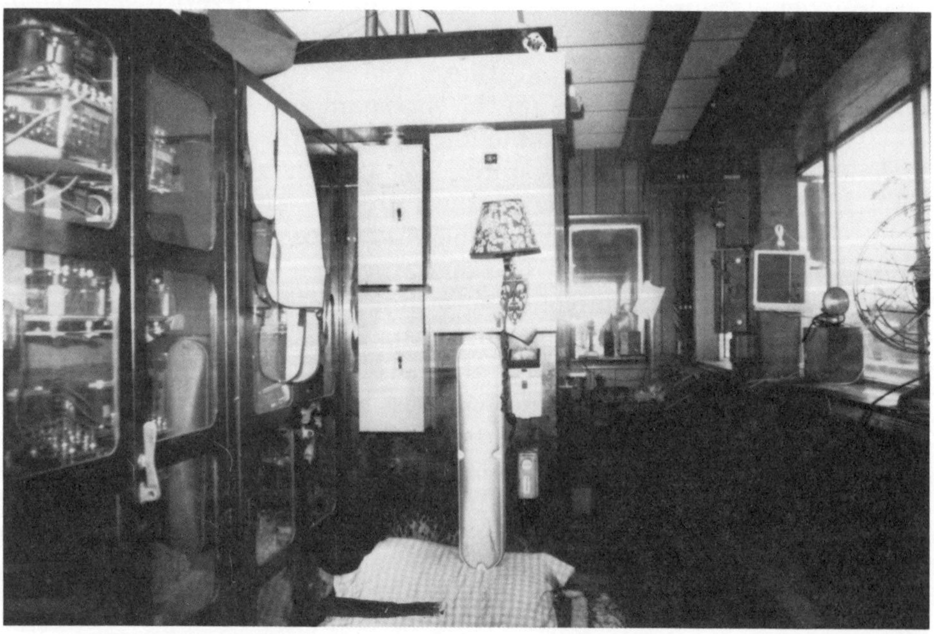

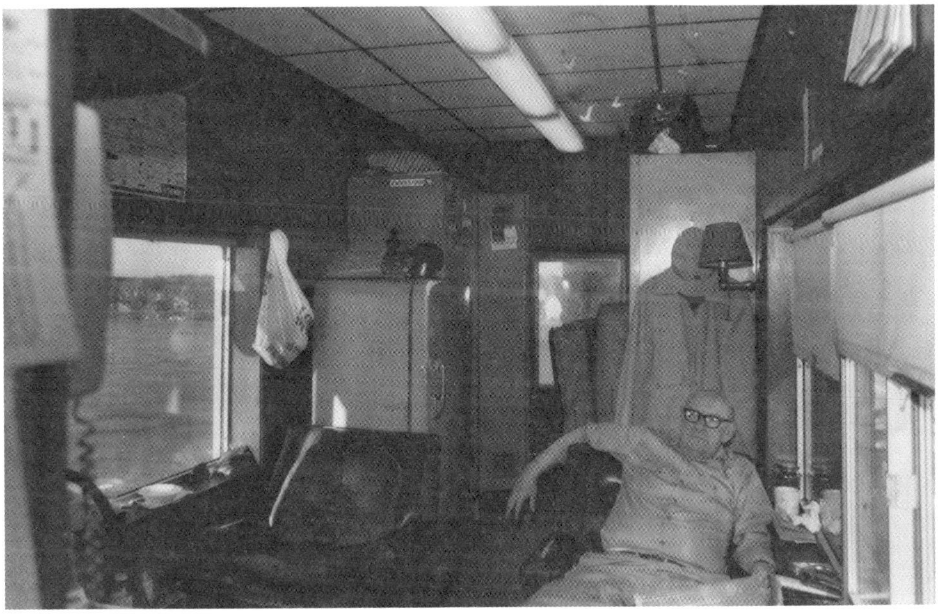

Interstate Bridge. Operators' house. Second story. James Howard, Operator. Photographer facing north. September 4, 1987. Photograph by author.

mately twenty-one feet above the highway deck, he pushes the bypass button; that continues to provide power until he lets it go—essentially, it is a "deadman's switch." A light comes on, and he has nine seconds to push the switch before the operation stops (the same sequence occurs at the top of the lift). The reason for this is to slow the span down before it gets to the bottom (or top, as the case may be). The operator then lowers it slowly to about two inches above the highway deck, stops, then seats it. Then, the tiller is turned to "off," the bell is shut off, the span gates go up, the land gates go up, the traffic lights change automatically, and the traffic starts flowing again. The gates are monitored by red and green lights on a panel over the window facing the roadway; these show when the north and south gates are open and closed, and when the cycle ends. Gauges behind the operator show if the span is out of skew to the north or south and, if so, how many inches (these were added some years ago; originally, the instrument consisted only of a horizontal needle). The operator then corrects the problem by turning a handle on the panel which supplies power to one end of the span or the other. Gauges on the panel show the kilowatts going to the lift motors and the synchronizing motors, so the span can be slowed if it starts going too fast (this is more likely to happen when lowering the span than when raising it). The span can

Interstate Bridge. Operators' house. Second story. Stairway at end of hall to left. Mirror above television set to see river traffic. September 4, 1987. Photographer facing south. Photograph by author.

be operated at about three inches out of skew; at five inches, the power shuts off. Throughout the operation, the gateman is located on the north side of the lift span with a 'walky-talky,' and can communicate with the operator to tell him if anything is out of order on that end (The Memorial Bridge differs from the Interstate in having two gatemen, due to the location of the operator's house, which is in the middle of the lift span over the roadway). On occasion, vessels do not signal at all, so it is up to the operator and gateman to watch for approaching vessels, although this is not specified in the regulations. For this purpose, mirrors are located in the operator's house, so that personnel can see the river even when seated upstairs.[56]

Although federal regulations specify that in the case of movable bridges over navigable waters, marine traffic has the right-of-way over land traffic, in practical terms the great increase in both types has led to the need to balance their requirements. Bridge operators complain of the need to open the lift span frequently in the midst of heavy highway traffic to allow small sailboats to pass. Some small sailboat skippers will not use the movable small craft span although they could—either they are unaware of it or they are afraid of it. On the other hand, commercial river traffic cannot be interrupted, due to legal and practical reasons. A lift may take as long as forty minutes if two ships pass through at the same time. In an effort to ease the problem, a movement to institute regularly-scheduled openings for small craft has gained headway within the last year or two. The project is favored by the Bridge Authority as a way of reducing highway traffic congestion and the number of bridge lifts.[57]

VIII *The Reorganization*

THE BONDS WITH WHICH THE BRIDGE was financed were paid off as of January 1, 1962, leaving the Authority with funds of approximately $500,000. The compact between the two states and the Act of Congress setting up the Authority had provided that no bridge could be built which would compete with the Interstate Bridge while the latter was operated as a toll bridge. Although it seems clear the Piscataqua River Bridge, opened in 1972 to carry I-95 between Portsmouth and Kittery, would fall under the provision, the Interstate Bridge bonds had been paid off and a maintenance trust fund built up when it was opened. The existence of this fund and the fact that the two states faced the imminent end of the federal interstate highway program (which had provided funds to build the interstate highway system, including I-95) lent urgency to their 1967 application for federal aid to build the I-95 bridge.

Under the terms of the Maine and New Hampshire statutes under which the Authority was set up, after paying off the bonds, it was required to accumulate a trust fund to provide for maintenance of the bridge. This had been done, with over $6,000,000 being accumulated by the time tolls were taken off the bridge at midnight of October 31, 1972. Although this was timed to correspond with the opening of the new bridge, the Authority believed that the income from the fund would be sufficient to maintain the Interstate Bridge and the bypass in perpetuity. During the period of 1972 to 1982, the book value of all funds held by the Authority was maintained at slightly over $6,000,000. The rate of return on these funds was about 6 percent a year (and somewhat more toward the end of this period, rising considerably in 1981-82) generating roughly $600,000, while the operating budget was around $250,000. Costs were rising in the 1970s, but not enough to threaten the maintenance of the facilities.

From November 1, 1972 to June 30, 1982, the Authority operated under a Plan of Procedure which designated the First National Bank of Boston as custodian of its "Permanent Investment Fund Principal" and set up a separate account entitled "Permanent Investment Fund Income" for the deposit of income collected on investments. The latter fund could be disbursed only upon written request of the Executive Secretary of the Authority. The bank was restricted to investing in U.S. Government obligations with a maturity of no more than fifteen years.[58]

This arrangement worked well. The Authority was able to live within its means until 1980, when it first noticed a drop in funds, due to the inflation beginning in the late 1970s. The income and expense curves began to come together, prompting the Authority to investigate means of increasing its revenue. For 1981-82, the operating budget was $374,978, and the capital improvement budget $680,000, making a total of over a million dollars. By contrast, the budgets for 1975-76 were $247,000 and $143,781.02, respectively, and the latter figure included $23,781.02 carried over from the year before. Although these figures do not tell the whole story (for example, capital improvement budgets fluctuate considerably and money is typically set aside over a period of years for projects such as painting), it is clear that costs rose enormously in the 1970s.[59]

The Authority tried to counteract this by increasing its revenues. On November 4, 1981, it instructed the First National Bank of Boston to sell certain U.S. Treasury Bonds and purchase higher yielding Treasury Notes (for example, on December 15, 1981, the Authority authorized the bank to sell $797,000 principal amount 7-7/8% Treasury Notes due May 15, 1986, held in the Permanent Investment Fund, and reinvest the proceeds in 13% Treasury Notes due November 15, 1990.[60]

It quickly became evident that the trust fund would run out within a relatively short period of time. To stave off the impending financial crisis, several alternatives were considered, such as reinstating the tolls. Considering tolls, the Authority commissioned a traffic survey by Wilbur Smith and Associates and consulted Attorney Frank Hancock regarding the legality of such a move. The survey considered one-way tolls, two-way tolls, and the optimum toll, which was discovered to be forty cents (The Forty-Cent Bridge?). However, the plan was not economically feasible. Furthermore, Attorney Hancock concluded that the Authority could not charge tolls for a purpose other than paying off a bond issue.

At this time, the Authority began seriously considering a reorganization which would allow the bridge and bypass to be maintained properly under the changed economic circumstances with the least possible financial impact on the citizens of the two states. The Authority had always managed to keep its funds intact, in effect operating the bridge cost-free for the states. However, during the period 1972 to 1982, operating expenses increased tremendously, mostly starting in 1977; from $233,000 in that year, operating costs nearly doubled to $440,000 in fiscal 1982, while trust fund income increased a lesser amount from $392,000 in fiscal 1976 to $491,000 in fiscal 1981.

Facing dissolution, the Authority signed an agreement with the First National Bank of Boston under which the Permanent Investment Fund assets were transferred to an Endowment Management Account, giving the bank more flexibility in managing the Authority's investments; annual income from this account as of June 1984 was estimated to be $611,338, which was a great improvement over previous years.[61]

The Authority then told the two states (in a friendly way) that if they didn't help it out financially, the entire burden of operating the bridge would be dumped in their laps.[62] The two states set up a Joint Task Force to study operating alternatives for the Authority, including placing the roadways on the Federal Aid System, to qualify for matching funds for major improvement projects. The Authority endorsed this concept at its meeting of December 20, 1983. At the May 18, 1984 meeting, John Petersen, Bridge Maintenance Engineer for New Hampshire, presented the rough draft of a Memorandum of Understanding between the Maine and New Hampshire Highway Departments for Operation and Maintenance of the Bridge after July 1, 1984. At a special meeting on October 26 of that year, the Task Force report was presented and discussed by Gedeon Picher of the Maine Department of Transportation and Robert Hogan of the New Hampshire Department of Public Works and Highways; Attorney Kim Visbaras of the Maine Department discussed the legal aspects. The report, entitled "Analysis of Operational Alternatives for the Maine-New Hampshire Interstate Bridge Authority," evaluated three alternatives. The first alternative was to continue the existing situation; the second was to dissolve the Authority in approximately ten years using the Authority's reserves in combination with federal aid to fund major improvements in facilities; and the third was to relegate control of the roads to the states while the Authority retained responsibility for the main river structure.

The Task Force recommended adoption of the third alternative, stating that "this plan provides the most comprehensive protection for the present Bridge Authority facilities, the most beneficial use of existing funds for those facilities, the least burdensome alternative pertaining to limited state and federal funds, and the maximum protection of the public interest for the citizens of Maine and New Hampshire." Adopted by the Authority at the October 26, 1984 meeting, that alternative leaves responsibility for the main river structure with the Authority, continuing its control of its current securities, while shifting all employees to the Maine Department of Transportation or the New Hampshire Department of Public Works and Highways. All work is to be done under a contractual arrangement with the two states, and roadway portions of the Authority have been transferred to the two states with the Authority financially responsible for capital improvements on the grade separation structures. The plan envisions continued solvency for the Authority until 2007 using reconstruction methods or until 2012 using rehabilitation for the main river structure and approaches, and it is specified that the Authority should give first priority to the main river structure. In practice, the Authority could continue to operate much longer, depending on prevailing economic conditions. The reorganization allows the Authority to use funds from the two states for 80 percent of the cost of approved capital improvement projects, with a portion of those funds coming from the federal government.[63]

The changes proposed in the plan based on the third alternative had to be approved by the legislatures of the two states and Congress. According to Robert Allard of the Bridge Authority, "The measure went through the Maine legislature 'lickety-split,' but met opposition in the New Hampshire House of Representatives from James Chandler, who felt that the state couldn't financially take on any more road. On the final vote, the measure passed by three votes--astounding when one considers that it was a 'non-issue'."

As for the intricacies of getting the bill through Congress, Allard said: "The bill was referred to committee and the chairman refused to allow it out for a vote because he wanted something from the Maine and New Hampshire legislators. Of course, they could have cared less, and said it would come out eventually. Well, it didn't come out at all...Because we were created by Congress, they had the final say. The first time it went through and it failed. Finally, it got through. Once we did that, we said we were home free. Not at all. There were the technicalities of transferring properties and people. Properties were simple. We found three pieces of property of no value to anybody, down near the new hospital (in Portsmouth). We approached the abuttors and invited them to bid. It worked out well. Then we got to the people. We have to keep our lift operators and gatemen together to operate the bridge. But the repair crew and maintenance...We said 'Where do you want to go?' Cans of worms! Each one wanted to go to a different state...retirement, wages...It's going to cost us $300,000 that the Bridge Authority has to pay the State of Maine for retirement benefits. To develop that number took about a year. Complex! My God! Complex and tedious beyond belief. We guaranteed everybody they would not suffer financially and would get a job they wanted, working for New Hampshire or Maine. The State of Maine had to create five jobs...the State of New Hampshire fitted in people through attrition...That was probably the most exciting part of my time...."[64]

Having been approved by the two state legislatures and the U.S. Congress, the plan finally went into effect, with February 2, 1986 fixed as the transfer date for employees and land. On the same day, the Authority deeded its property in Maine and New Hampshire to the two states respectively. The Compact Amendment between the two states became effective on December 19, 1985, with the provisions that the Maine Department of Transportation is to perform winter maintenance on that portion of the main river structure lying within the State of Maine (0.4742 lane miles), that the Maine Department of Transportation may employ independent consultants or contractors, and that the Authority agrees to reimburse the MDOT for costs incurred under the above two items. An agreement between the Authority and the New Hampshire Department of Public Works and Highways provides that the State of New Hampshire is to perform winter maintenance on that part of the main river structure lying within the State of New Hampshire (0.6496 lane miles), that it is to perform physical maintenance of the roadway portion for the entire river structure, that it shall coordinate and administer the operation of the main river structure including the drawspan and gates, coordinating with appropriate utilities and initially paying for said utilities, and that it may retain or employ independent consultants or contractors.[65]

Interstate Bridge. 1940 dedication plaque (one of two identical plaques at opposite ends of the bridge), on east side of handrail pier, north bridge abutment, west side of road. November 1987. Photograph by author.

IX *The Maine-New Hampshire Interstate Bridge Authority*

THE HISTORY OF THE Interstate Bridge Authority and the bridge which it built, and has cared for these many years, is part of the rich history of Portsmouth and Kittery, and the two states of which they are a part. Renamed the Sarah Mildred Long Bridge on June 5, 1987, in honor of the mainstay of the Authority since its formation ("she went in there with a typewriter and a portable table"), the bridge has been the focus of a remarkable group of dedicated people over the past fifty years. As Authority member Mrs. Rita Perry commented, "The Authority and the bridge will probably last one as long as the other...I have a feeling of proprietorship." Dana F. Connors, Chairman of the Authority and Maine Highway Commissioner, became a member in 1984 and soon thereafter assumed the chairmanship. He sees the Authority as "An organization that brings me to many involvements," and has "a special fondness for this one because of the interest and dedication that everyone brings to (it)...Everybody pitches in... The captain of the team is Mildred Long...."[66]

The following is a list of the many dedicated members of the Interstate Bridge Authority.[67] Unless otherwise noted, the dates indicate when individuals assumed membership; * indicates state highway commissioners:

Representing Maine	Representing New Hampshire
6/03/37	
HINMAN, WM H., Skowhegan; V.Chairman	*EVERETT, FREDERIC E., Concord; Chairman

COLE, HOLLIS B.,
Kittery; Treasurer

*THURSTON, PAUL C.,
Bethel

BROOKS, FRANK E.,
Portsmouth; Clerk

POWERS, JAMES J.,
Manchester

5/04/39
*WOODMAN, STILLMAN E., Machias; replaced Mr. Thurston

6/09/42
SAYWARD, JOSEPH T., Kennebunk; replaced Mr. Hinman

11/05/48
*MORTON, LLOYD B., Farmington; replaced Mr. Woodman

9/06/49
ELWELL, EDWARD,
West Buxton; replaced Mr. Cole

*MERRILL, FRANK D.,
Concord; replaced Mr. Everett

4/17/52 to 8/15/52
MORTON, JOHN O., Concord; served on Authority while General Merrill was on active duty in Army

5/26/53
JARVIS, ANDREW,
Portsmouth; replaced Mr. Brooks

2/26/54
*STEVENS, DAVID H., Augusta; replaced Mr. Morton

DOW, HAROLD L., Eliot; replaced Mr. Sayward

12/31/55
HOAR, ALLEN, York; replaced Mr. Elwell

*MORTON, JOHN O., Concord; replaced General Merrill

3/16/62
HUSSEY, PHILIP W., Jr., Kennebunkport; replaced Mr. Hoar

2/25/65
KEENAN, JAMES P.,
Portsmouth; replaced Mr. Jarvis

2/26/68
*WHITAKER, ROBERT H.,
Concord; replaced Mr. Morton

2/19/69
STAPLES, ROGER W.,
Eliot; replaced Mr. Hussey

LANDRY, NORMAN R.,
Portsmouth; replaced
Mr. Keenan

9/22/70
ALLARD, ROBERT A.,
Portsmouth; replaced Mr.
Landry

2/18/73
Mr. STEVENS, Hallowell;
retired from Highway Commission

4/22/74
*MALLAR, ROGER L., Augusta;
replaced Mr. Stevens

STEVENS, DAVID H., Hallowell;
replaced Mr. Dow

11/29/75
WHITAKER, ROBERT H.,
Concord; retired as N.H. High
way Commissioner

2/21/76
DR. POWERS died

3/26/76
*CLEMENTS, JOHN A.,
Concord; replaced
Mr. Whitaker

9/17/76
KENNEDY, WILLIAM R.,
Hampton; replaced Dr. Powers

10/15/79
SILVERMAN, HAROLD J.,
Portsmouth; replaced
Mr. Kennedy

12/28/79
MALLAR, ROGER L., Augusta;
resigned as Highway Commissioner

6/27/80
*CAMPBELL, GEORGE N., Jr., Augusta; replaced Mr. Mallar

9/19/80
PERRY, RITA A., Kittery Point; replaced Mr. Stevens

12/02/83
CAMPBELL, GEORGE N., Jr., Augusta; resigned as Highway Commissioner

3/09/84
*CONNORS, DANA F., Augusta; replaced Mr. Campbell

12/31/84
CLEMENTS, JOHN A., Concord; resigned as N.H. Highway Commissioner

3/01/85
*CHANDLER, JOHN P., Concord; replaced Mr. Clements

1/07/86
*STICKNEY, WALLACE F., Concord; replaced Mr. Chandler

Membership on 11/20/87:

*CONNORS, DANA F., Augusta

STAPLES, ROGER W., Eliot
Treasurer
PERRY, RITA A., Kittery Point

STICKNEY, WALLACE F., Concord; V. Chairman

ALLARD, ROBERT A., Portsmouth; Asst.Treas.
SILVERMAN, HAROLD J., Portsmouth; Clerk

CHAIRMEN OF THE AUTHORITY

Everett, Frederic E.	1937-1949
Morton, Lloyd B.	1949-1954
Merrill, Frank D.	1954-1955
Stevens, David H.	1955-1973

Whitaker, Robert H.	1973-1976
Mallar, Roger L.	1976-1979
Clements, John A.	1979-1984
Connors, Dana F.	1984-

The Chairman of the Authority is always the Highway Commissioner of Maine or New Hampshire, with the Vice Chairman being the Commissioner from the other state. The Highway Commissioners automatically become members of the Authority. Other members are appointed by the governor and confirmed by the council of each state, and serve for terms of six years in Maine and four years in New Hampshire. Usually, there is one member from the minority political party in each state. Officers are elected at the Annual Meeting by the members of the Authority. Two members from each state must be present to constitute a quorum. Meetings are ordinarily held quarterly, but special meetings may be called as required.

Of the original members of the Authority, Frederic E. Everett

Rededication of the Sarah Mildred Long Bridge. Miss Long has just cut the ribbon! Left to right: Robert H. Whitaker, former member of the Authority; Mary Keenan, Mayor of Portsmouth, N.H.; Harold J. Silverman, member of the Authority; Robert Allard, member of the Authority; the Reverend Michael Quirk; Miss Long; Ruth Griffin, Executive Councilor; Dana Conners, Maine State Commissioner of Transportation; Roger Staples, member of the Authority; Eric Strahl, Kittery, Maine, Town Manager; Wallace Stickney, New Hampshire Commissioner of Transportation. June 5, 1987. Photograph courtesy of James Bartlett; Public Information Office; N.H. Department of Transportation.

of Concord was the New Hampshire Highway Commissioner, Frank E. Brooks of Portsmouth, N.H., owned a Ford automobile dealership, and Dr. James J. Powers of Manchester, N.H., was a physician; Paul C. Thurston of Bethel was the Maine Highway Commissioner, William H. Hinman of Skowhegan, Me., had a contracting business, and Judge Hollis B. Cole of Kittery, Me., was an attorney and Judge of the Kittery District Court.[68] As an historical footnote, General Frank D. Merrill, New Hampshire Highway Commissioner and a chairman of the Authority, became famous in World War II as the leader of Merrill's Marauders in Burma.

Over the years, members of the Authority can generally be placed in one of two groups—transportation professionals, and individuals having a connection with one major party or the other and an interest in public service.[69] However, the Authority can hardly be described as political and the balance between professionals and nonprofessionals has been an asset.

Of the executive secretaries and maintenance superintendents, Wallace Purrington came from Concord, New Hampshire, during construction of the bridge. He served as Clerk of the Works, and stayed on as Executive Secretary of the Authority, retiring c.1955. Worth Pridham was the first Superintendent of Maintenance, but did not stay long; he was followed by Raymond E. (Johnny) Morrow. Mr. Morrow followed Mr. Purrington as Executive Secretary, while continuing as Superintendent of Maintenance; he retired in 1975. At that time, Hiram A. Young, Jr. became Superintendent of Maintenance, having worked for the Authority since 1952, and Miss Mildred Long became Executive Secretary, having worked for the Authority since 1937. Officially retired since 1984, Miss Long continues on a consulting basis; she remains the *de facto* Executive Secretary of the Authority. Under the 1986 reorganization, the Authority no longer has any employees. Ralph Flynn was the first Chief Operator, and the other operators when the bridge opened in 1940 were Harry Seawards, Gordon Jameson and James Abel. There were no gate tenders until increased traffic after World War II necessitated their hiring.[70]

The Sarah Mildred Long Bridge has been fortunate to have an extraordinarily bright group of people interested in it over the years, beginning with its planning, financing and construction. While the Maine-New Hampshire Interstate Bridge Authority has the ultimate responsibility for the Bridge, its effectiveness is largely dependent on having good advice from experts (whether it be in engineering, financial or other areas), good operational management through the Executive Secretary or equivalent, and reliable employees—operators, gatemen, tolltakers and maintenance crew members. The Authority was lucky to find an engineering firm which had the breadth of knowledge to advise the board on virtually every aspect of bridge construction, operation and maintenance, and has taken a personal interest in the structure over the years. Likewise, the bridge has benefitted from the expertise of the transportation departments of Maine and New Hampshire—a really dedicated group of people. It has also had the advantage of great continuity of personnel.

The Long Bridge connects the wealth of local history, represented by its predecessor and the two communities joined by it, with the larger worlds of vertical lift bridge engineering and national economic history—it was built at the end of the Great Depression and was greatly affected by fuel rationing in World War II, the prosperity of the 1950s and 1960s, the building of the Piscataqua River Bridge as part of the interstate highway program, and the inflation culminating around 1980. In the words of Ernest Searles, Maintenance Engineer for the New Hampshire Department of Transportation, "This is a continuing story."

X *Voices of the Bridge*

BASED ON NUMEROUS INTERVIEWS, the following attempts to give an idea of various aspects of the bridge operation. It is fortunate that there are still people around who were involved with the bridge from the beginning. One of these is James Abel, who talked about his experiences as a bridge operator for thirty-one years, starting when the bridge opened. His recollection of its operation as part of the Eastern Division of the Boston and Maine Railroad is worth preserving, because he is one of a select number who can recall the details:

> The (railroad) tunnel went to Butler's Crossing, in Kittery, where 103 goes to Eliot. You used to go from there...there's a church; it came out at (the later Route) 236 at Fernald's Crossing. We used to control the signals in the operator's house...(extending) in Portsmouth from Emery's Crossing near Barberry Lane, and in Kittery...from Fernald Road. We had four levers—the gates on the bridge and control of the trains three miles each side of the bridge. Once the train got inside those points, the bridge couldn't be opened until it crossed to the other side; there was a wait of three minutes to change the railroad lights from green to red and the ship channel lights from red to green, and the railroad signals were 'High yellow' at either end—approximately three miles out, two reds near the bridge and three reds at either end of the bridge. If the gates wouldn't go down or the lights wouldn't work, you couldn't lift the bridge...It was an automatic relay; it turned red if the train stopped in Portsmouth. It took about 3-1/2 minutes to clear. Boats had the right of way over trains.[71]

Interstate Bridge. Using the lift span to retrieve an automobile which had gone off the bridge. Barge approaching to carry it to the Navy Yard in Kittery. A woman died in the accident; her body was still in the car. 1949. Photograph by Paul E. Marston, courtesy of Roger Staples. Photographer facing east.

Clarence Foley, an operator, recently talked about his past fourteen years on the job, a second career for him (he had been in retail work, his children were grown and he wanted some time to himself). When he started as an operator fourteen years ago, he was checked out by Robert Williams, one of the former operators, to operate the lift span. He started the sequence, having looked to see that no traffic was approaching the bridge. Suddenly, out of nowhere, a car came flying onto the bridge just as the gates went down ("It was going ninety miles an hour, if it was going one"). Foley tried to stop the gates, but it was too late. The car, a Cadillac, hit the span gate, breaking its windshield, and came to a stop. He went out of the operators' house, looked in the car and asked the occupants (an older man and a young woman) if they were alright, and why hadn't they stopped? They were on the way to the airport in Portland; yes, they were alright. Williams, an old Mainer, didn't say a word. The next day,

Foley was certified to operate the span.[72]

Likewise, James Abel's recollection of a similar experience while he was an operator shows that working on the bridge is not all routine:

> I was there one night when this car pulled up at the gate and stopped. The bridge was up for the tugs. It was in the winter; it was pretty cold. When I was writing in the logbook, I looked up and saw the woman standing by the draw. I tried to talk her out of it and grabbed her wrist. The woman jumped off. I radioed to the tugs. They picked her up. She lived. There were cars that went off. The Navy Yard pulled one out with a crane because it was laying on cables...We had a lot of 'em go through the gates. They'd go right through (those) wooden gates. We could see them coming and wouldn't lift."

Robert Allard, who obviously enjoys his work as a member of the Authority, described the experience in this way:

> I was appointed through Bob Whalen, friend and competitor (Whalen Mayflower Moving Company; he was a member of the Governor's Council)...He called and said 'I'm going to appoint you to the I.B.A.' I didn't know if I was managing all the bridges in New Hampshire or what. I got a notice to appear at a meeting at the Rockingham Hotel. I was twenty-nine. Young. The members of the Authority were all very mature. Roger Staples was in his fifties—the youngest. Dr. Powers was in his seventies. I walked into the meeting. I still didn't know where the bridge was...I walk into this dark ballroom. The first person I walk into is Dr. Powers. He thought I was my father...Couldn't relate...Too much of a gap. I tried to explain. That was a riot. Here's Dr. Powers, one of the originators. Then I met David Stevens. Probably the most astute intellectual I ever served with. I figured they would run the meeting around me and I'd try to pick it up...Mr. Stevens spent probably three-quarters of an hour in reviewing the history of the bridge. He talked about the small boat channel... He really ran the meetings. The guiding light of the Commission...He was always right. He knew what we could spend, and Mr. Whitaker (from New Hampshire)... When Dr. Powers died, we made sure the treasurer was really local. So we made Roger the treasurer, and I was the assistant. We met quarterly and the meetings were long and involved, but we accomplished a lot. We always had a lot of lobster sandwiches. Miss Long would bring them in. Early on, we met at Dr. Powers' home on the beach. We met then at the Rockingham, then moved over to Maine and had

> meetings at the office (on Post Road in Kittery).
>
> We were paid twenty dollars per meeting. I think the treasurer gets ten dollars for signing checks. The commissioners don't get any money. The other four do. The most I ever made in a year was $180. I have been appointed five times. It's an interesting appointment. I'm a Republican. I have been appointed by Democrats and Republicans. Hugh Gallen (Democratic Governor of New Hampshire) appointed me. I work hard for the party and the Governor knows that. It was out front… Thomson (Republican Governor of New Hampshire) appointed me and I wasn't on his team…
>
> We're making about a half million dollars a year on investments. We're close to hitting principal. These are very conservative investments. They haven't been wild. Within a very short time, we're going to run out of interest and have to touch principal. We tried to put the tolls back on (c.1982). We discovered there was no toll capable of generating money to protect the principal. Graphed how it would affect traffic flow. Forty cents was the best point for traffic flow, but the bottom line was so small, it was not worth the effort. We took the tolls off because the I-95 bridge opened. We would've left them on indefinitely…"[73]

The comments of David Stevens, former Chairman of the Bridge Authority and recently retired from the Maine Turnpike Commission, also shed light on the political considerations of running the bridge and bypass, as well as the effort to restructure the Authority's responsibilities in the 1970s (before the reorganization plan was adopted):

> We attempted to arrange for the approaches to go to the two states. The New Hampshire approach included the stretch from the bridge to Route 1. Portsmouth was forever pressuring the Authority to construct various additions to that section of road. I suggested the Authority should divest itself of that stretch, to the State. John Morton, Highway Commissioner of New Hampshire, was very aware of the problems of carrying out the job; he said that if New Hampshire took over that stretch, they would be under intense pressure from the City to do things the City wanted done. John realized he would have trouble standing up to the political pressure of Portsmouth, so he resisted any attempt of the Authority to divest itself of that stretch of road. I wanted to divest the Authority of both approaches. The difficulty was that the Authority had accumulated a sizeable trust fund. We discussed breaking up the fund to give New Hampshire a bigger piece because there was more road in New Hampshire.

David H. Stevens, former member and chairman of the Authority, retired in July 1987 as Executive Director and Secretary-Treasurer of the Maine Turnpike Authority, at the age of 80. Photograph courtesy of Guy Gannett Publishing Co.; Portland, Maine. Published in the Portland, Maine, Press Herald on July 28 and August 13, 1987. A small town native frequently described as "one of the most brilliant men I've encountered," Mr. Stevens can tell the listener how to build a bridge or turnpike, from planning, through legislation and financing, to completion.

> John Morton and his successor, Robert Whitaker, both unofficially and quietly admitted their inability to handle the state legislature in New Hampshire. Their concern was that it was difficult to control the legislature due to its size, and New Hampshire is very prudent. They were afraid that in their zeal these people would become aware of the trust fund and would 'grab it and go,' to avoid new taxes, and then try to maintain the bridge out of current funds.

One might add that the political ability of David Stevens is legendary, and that his ideas may be confidently expected to influence Maine transportation policies for years to come (He retired in 1987 at the age of 82).[74]

As finally carried out, the reorganization of the Authority says

Raymond E. (Johnny) Morrow of Kittery, Maine. Executive Director and Maintenance Supervisor for the Bridge Authority. C.1970. Photograph courtesy of his sisters.

much about the current preference for repair versus replacement of transportation facilities. As explained by Wallace F. Stickney, New Hampshire Commissioner of Transportation, attitudes have changed since the "glory days of the Interstate system," when the mentality was to replace old structures. Now there is a lot more rehabilitation and less total rebuilding of bridges. The Hinsdale-Brattleboro (New Hampshire-Vermont) bridge is an example. Rehabilitation was chosen over rebuilding because the cost would be approximately $1.8 million, versus $6.4 million. As for the Interstate Bridge, it should be functional for at least another 20-25 years; even if it were a new bridge now, it would have to be redecked during that time.[75]

Beyond all of the other experiences of building and maintaining the bridge is a story of friendship. In August 1987, David Stevens recalled serving on the Authority with Andrew Jarvis, Portsmouth restaurant operator and former owner of the Rockingham Hotel, where the Authority often met; he remembered Mr. Jarvis as a fine gentleman, a self-made man with the Old World values of family, friendliness and compassion—"Old-world courtliness."

Rita Perry, who was appointed to the Authority in 1980 by Governor Joseph Brennan of Maine, a Democrat, suggested the continuity of service and friendship which characterizes the history of the bridge over a half-century:

> It was a political appointment, but I was very familiar with the bridge. While I was on the paper (the *Portsmouth Herald*), the High Level Bridge was proposed. George Varney was active with the Authority; later, he was a district judge here (in Kittery). I got through at the paper in 1967, and the High Level opened in 1972. Varney was one of the early ones (associated with the Interstate Bridge project). When I was appointed, I got two calls—Frank Kaufman, a local civic-minded man; he wanted the appointment and congratulated me. George Varney called; he had just retired as a lawyer and was interested in it. The calls meant a lot to me. He (George Varney) was a friend of Dave Stevens. There was a group who became friends. All civic-minded people. George Varney's beagle fathered a litter of ours—Ladybelle Mike. All the beagles around here were related. They (the men on the Authority) used to play poker at the old Authority building on Post Road. They were all woodsmen. They didn't care about shooting rabbits, they just liked to get out there. Good men. The office is now a barber shop.

Conclusion

That is the story of the Dime Bridge—a story of civic pride, engineering skill, devotion to duty and friendship. More than can be put into words. The rededication of the Maine-New Hampshire Interstate Bridge as the Sarah Mildred Long Bridge recognizes this devotion and presents a challenge to future generations to maintain a noble tradition of public service.

Notes

1. Chandler B. Cobb, Kittery Point, Me., interview, October 20, 1987.

2. Raymond A. Brighton, *They Came to Fish,* 119 and 127. "Making Final Inspection of Railroad Bridge," *Portsmouth Herald* April 18, 1939. Hiram Young, interviews, August 13 and October 8, 1987. James Abel, interview, October 9, 1987. *Portsmouth Herald* July 19, 25 and 26, 1904, and September 11, 1939.

3. Robert Williams, interview, November 27, 1987.

4. Roger Staples and Leland Riley Jr., interviews, October 29, 1987.

5. Norman Clark, interview, November 25, 1987. Brighton, *They Came to Fish*, II, 334, n6.

6. Kennard Palfrey, interview, November 27, 1987. *Portsmouth Herald*, July 24, 1905. The Spinneys telephoned for an ambulance and assisted the injured passengers.

7. *Portsmouth Herald* , Nov. 8, 1940, p.12.

8. *Portsmouth Herald* , July 20, 1935.

9. I.B.A. Minutes, June 3, 1937.

10. Ibid.

11. Reasons for selection reported by Miss Mildred Long, interview, July 25, 1988. The I.B.A. Minutes are not clear on this point.

12. I.B.A. Minutes, February 19, 1938.

13. I.B.A. Minutes, September 14, 1938. Coverdale & Colpitts, Consulting Engineers, 120 Wall St., New York: Letter to Bridge Authority dated September 13, 1938.

14. Tripp & Co. Prospectus, February 1, 1955.

15. I.B.A. Minutes, November 1 and 3, 1938.

16. *Portsmouth Herald*, November 8 and 9, 1940.

17. George A. Hool and W.S. Kinne, 165-171. David Plowden, 188f. Harrington and Cortelyou, historical brochure.
Published works by Waddell and Harrington include the following, all of them available in the University of New Hampshire Library, Durham, N.H.:
J.A.L. Waddell, *Bridge Engineering*, 2 vols. (N.Y.: John Wiley & Sons, 1925), see Introduction). Dr. Waddell also published *Memoirs and Addresses of Two Decades* (Edited by Frank W. Skinner, Easton, Pa.: Mack Printing Co., 1928), *Principal Professional Papers* (Edited by J.L. Harrington, New York, 1905), and *Addresses to Engineering Students* (Edited by Waddell and J.L. Harrington, 2nd. ed., Kansas City, Mo.: Waddell and Harrington, 1912).

18. N.H. Historic Bridge Inventory: Portsmouth Bridge No. 247/084, prepared for N.H. Dept. of Public Works and Highways by Sverdrup & Parcel and Associates, Inc., Boston, Sept. 1982, 3. Norman Y. Holt, interview, November 19, 1987.

19. Otis Ellis Hovey, 26f.

20. Plowden, 189. Hool and Kinne, 179b, 179c and fig.17. These rectangular braced towers are cased-in and narrowed at the top, where they house the drums and other machinery. The weight of the lift span given in 179c is erroneous: Hiram Young, interview:

August 13, 1987.

21. Interstate Bridge: General Plan and Elevation, Harrington & Cortelyou, Consulting Engineers, Jan. 17, Feb. 10 and 16, 1939. Tower: Sheet no. 34, May 4, 1939.

22. Hiram Young, Supervisor of Maintenance, interview, Oct. 8, 1987. Harrington & Cortelyou, Consulting Engineers, Plans for Maine-New Hampshire Interstate Bridge, Sheet 36, May 25, 1939. I.B.A. Files, Bridge Maintenance Office, Kearsarge Way, Portsmouth, N.H.

23. Miss Mildred Long, various interviews, summer and fall 1987. Edwin M. Newman, letter, Oct. 25, 1987, and telephone interview Oct. 1, 1987. Willis Grinstead, interview at I.B.A. Meeting, Oct. 9, 1987.

24. Edwin M. Newman, letter, Oct. 25, 1987.

25. *Portsmouth Herald*, Feb.28, 1939.

26. *Portsmouth Herald*, April 20, 1939. H.D. Peoples, telephone interview, November 14, 1987.

27. *Portsmouth Herald*, June 15, 1939.

28. H.D. Peoples, taped recollections, Nov. 17, 1987 (The following construction outline is based on his information). *Portsmouth Herald*, July 13, July 22 and Aug. 3, 1939.

29. *Portsmouth Herald*, Feb. 1 and April 27, 1940.

30. *Portsmouth Herald*, July 13, 1939. I.B.A. Meeting, June 6, 1940.

31. *Portsmouth Herald*, Aug. 5, 1940.

32. *Portsmouth Herald*, Nov. 8, 1940.

33. *Portsmouth Herald*, November 9, 1940. The *Union*, Manchester, N.H., Nov. 9, 1940. The *Boston Globe*, Nov. 9, 1940. For earlier bridges and ferries, see *Foster's Daily Democrat,* Dover, N.H., Sept. 18, 1971, p.11 (this includes information on the Piscataqua River Bridge then being built). The reader should also be aware of the Carlton Bridge, a combined highway and railroad bridge

between Bath and Woolwich, Maine, located on U.S. Route 1, this 3,100-foot bridge was built in 1927 by the McClintic-Marshall Company and the Foundation Company of New York, and was designed by the firm of J.A.L. Waddell. The lift span is a 234' Warren truss; it provides a vertical clearance of 135'. Information on the Carlton Bridge can be found in the *State of Maine Department of Transportation Inspection Report for the Carlton Bridge Lift Span* (July 1986, by Hardesty & Hanover, Consulting Engineers, successor firm to J.A.L. Waddell).

34. The reader will notice that there are four Maine members of the Authority listed on the dedicatory plaque. The Authority was originally formed on June 3, 1937; on May 4, 1939, Paul C. Thurston replaced Stillman E. Woodman, so both served on the Authority while the bridge was being built. Otherwise, all of the original members of the Authority remained from its inception through the dedication and for several years thereafter. That explains why seven members are listed, which was never actually the case. There have always been three members from each state.

35. James Abel, retired operator, interview, Oct. 10, 1987.

36. Ibid.

37. Harry L. Laderbush, interview, November 4, 1987. Norman Clark, interview, November 18, 1987 (he purchased land from Mr. Smith to build his house on Woodbury Avenue). Miss Mildred Long, interview, October 27, 1987.

38. I.B.A. Minutes, Dec. 5, 1951. Leland Riley, interview, Oct. 29, 1987.

39. Chandler B. Cobb, Kittery, Me., interview, October 16, 1987. U.S.G.S. Topographic Map, SE/4 Dover 15' Quadrangle, 1956. *Boston & Maine Railroad Newsletter* for October 1952: "Old 'P.S.&P.' is Gone" (Newsletter courtesy of Mr. Cobb).

40. Tripp & Co. Prospectus, Feb. 1, 1955.

41. *Portsmouth Herald* , May 6 and Sept. 29, 1964; Feb. 5 and 10, 1965; Jan. 11, 1967.

42. Ernest Searles, Maintenance Engineer, N.H. Dept. of Public Works and Highways, interviews, August 26 and December 4, 1987.

43. *Portsmouth Herald*, Jan. 11, 1967; Sept. 9, 1965, Dec. 18, 1965; and April 17, 1965.

44. John A. Herr report, October 8, 1968. Ernest Searles, interview, December 4, 1987. Harrington and Cortelyou, drawings entitled "Rope Adjustments," revised March 16, 1977. I.B.A. Minutes, September 17, 1976, and March 18, 1977, re. efforts of David Dickey of the N.H. Department of Public Works and Highways to solve the vibration problem.

45. Letter: Raymond E. Morrow, Executive Secretary, Bridge Authority to A.R. Sirois, Bridge Maintenance Engineer, Maine State Highway Commission, January 18, 1971. *Seventeenth Biennial Report of the Maine-New Hampshire Interstate Bridge Authority*, March 15, 1973, pp.2f. I.B.A. Minutes for February 17, 1972, and March 17, 1972.

46. *Portsmouth Herald*, May 25, 1971. Memorandum: Bernard H. Langley, Assistant Chief Engineer, N.H. Department of Public Works and Highways, to R.H. Whitaker, Commissioner, "Collision with NH Tower Pier by Corps of Engineers Hopper Dredge," May 27, 1971. Capt. Richard Holt, harbor pilot, interview, October 8, 1987. Hiram Young, interview, December 4, 1987.

47. *I.B.A. Seventeenth Biennial Report,* March 15, 1973, p.3. The report gives the date as June 8, 1971, but one assumes that this is the same accident, and that it actually occurred on the morning of June 9, giving Portsmouth and Kittery a taste of Boston-type traffic. *Portsmouth Herald*, June 9, 1971. Harrington & Cortelyou, consulting engineers, letter to David H. Stevens, Chairman, Maine-New Hampshire Interstate Bridge Authority, June 25, 1971.

48. F.M. Cortelyou, Jr., letter to Robert H. Whitaker, Chairman of Interstate Bridge Authority, June 18, 1974. *Portsmouth Herald*, September 29, 1975; April 7, 1976. Hiram Young, interview, October 8, 1987.

49. I.B.A. Minutes, April 1, 1982. Ernest Searles, interview, Aug. 26, 1987.

50. I.B.A. Logbook, Jan. 22 and 23, 1987.

51. James Abel, loc.cit.

52. Historical perspective on the issue is provided by a letter dated April 27, 1940, from Max L. Wilder, Bridge Engineer with the Maine Highway Department, to Wallace F. Purrington, Executive Secretary of the Authority: "In answer to your letter of April 26th, the State does not carry insurance on any of its toll bridges except, on three bridges, there is protection against interior robbery and messenger holdup, thus taking care of the money from the time it is received until it is deposited. The amounts vary with the probable cash on hand, being $5,000.00 for the Carlton Bridge, $2,500.00 for the Waldo-Hancock Bridge, and $1,000.00 for the Deer Isle-Sedgwick Bridge." As noted elsewhere, war risk insurance carried by the Authority is discussed in a letter dated April 6, 1944, from Mr. Purrington to Mr. Wilder.

53. Captain Richard Holt, interviews, Oct. 6 and 23, 1987. *Portsmouth Herald*, Oct. 11, 1984.

54. Ibid.

55. Ibid.

56. Clarence Foley, operator, interview, Oct. 27, 1987. James Howard, operator, interview, Aug. 20, 1987.

57. Charles Starkey, operator, interview, Oct. 15, 1987.

58. "Background Information for Analysis of Operational Alternatives for the Maine-New Hampshire Interstate Bridge Authority, Maine Dept. of Transportation and New Hampshire Dept. of Public Works and Highways," October 1984, by Robert Hogan of the New Hampshire Dept.

59. I.B.A. Minutes, June 20, 1975; June 27, 1980.

60. I.B.A. Minutes, Nov. 4 and Dec. 15, 1981.

61. I.B.A. Minutes, June 23, 1982.

62. Robert Hogan, "Background Information" for Operational Alternatives.

63. Roger Staples, interview, Nov. 13, 1987.

64. Robert Allard, interview, Oct. 14, 1987.

65. I.B.A. Minutes, Jan. 3, 1986, deeds and other documents attached.

66. Rita A. Perry, interview, Oct. 15, 1987. Dana F. Connors, interview, Sept. 2, 1987.

67. List prepared from I.B.A. Minutes by Miss Mildred Long.

68. Miss Mildred Long, interview, Oct. 27, 1987.

69. Rita A. Perry, loc.cit.

70. James Abel, loc.cit.

71. Ibid.

72. Clarence Foley, operator, interview, Oct. 27, 1987.

73. Robert Allard, loc.cit.

74. David H. Stevens, interview, Aug. 15, 1987. Concerning Mr. Stevens, see "Road trips end for retired turnpike director," in *Portland*, Maine, *Press Herald* for July 28, 1987, and "Turnpike project faces rocky road," in *Portland Press Herald* for Aug. 13, 1987.

75. Wallace F. Stickney, interview, Aug. 25, 1987.

An Annotated Bibliography

Publications, Letters and Reports

Barnard, Everett B., Bridge Maintenance Engineer, Maine Department of Transportation. Letter to author regarding date of Kittery Traffic Circle and Route 236 Underpass. December 1, 1987.

Brighton, Raymond A. *They Came to Fish*. Rev. ed. 2 vols. Dover, N.H.: Randall/Winebaum Enterprises, 1979. See I, p. 292, and II, pp. 175 and 375, regarding the Old Railroad Bridge (predecessor to the Interstate Bridge).

Cortelyou, F.M., Jr. Letter to Robert H. Whitaker, Chairman of Maine-New Hampshire Interstate Bridge Authority. June 18, 1974.

Coverdale & Colpitts, Consulting Engineers. Letter to Bridge Authority. September 13, 1938.

Hardesty & Hanover, Consulting Engineers. *State of Maine Inspection Report for the Carlton Bridge Lift Span.* N.Y.: July 1986.

Harrington and Cortelyou, Consulting Engineers. Letter to David H. Stevens, Chairman, Bridge Authority. June 25, 1971.
—with Ayres Associates, Consulting Engineers. "Maine-New Hampshire Interstate Bridge Authority, Piscataqua River Bridge; Report on Bridge Inspection and Recommendations for Repairs and Improvements." December 31, 1986. Note to the reader: "Piscataqua River Bridge" can also refer to the I-95 (High Level)

bridge opened in 1972—that is its official name; therefore, be aware of the context in which used.
—*Movable Bridges*, with enclosure, “History and Organization.” N.D.
—Plans: Maine-New Hampshire Interstate Bridge Authority Piscataqua River Bridge, Portsmouth, N.H. to Kittery, Me., P.W.A. Project No. N.H. 1037-F., Maine 1088-F., January 17, 1939, with subsequent revisions, and record revisions by the firm (F.M. Cortelyou) August 17, 1945.
—Telegram to Maine State Highway Department re. Public Works Administration Docket. December 15, 1938.

Herr, John A. Report concerning vibration problem. October 8, 1968.

Hogan, Robert, New Hampshire Department of Public Works and Highways. “Background Information for Analysis of Operational Alternatives for the Maine-New Hampshire Interstate Bridge Authority.” Maine Department of Transportation and New Hampshire Department of Public Works and Highways. October 1984.

Hool, George A., and W.S. Kinne, eds. *Movable and Long-Span Steel Bridges*. N.Y. and London: McGraw-Hill, 1943.

Hovey, Otis Ellis. *Movable Bridges*. N.Y.: John Wiley & Sons, 1926.

Langley, Bernard H., Assistant Chief Engineer, N.H. Department of Public Works and Highways. Letter to R.H. Whitaker, N.H. Commissioner of Transportation. May 27, 1971.

Lockwood, Harold J., Acting State Director, Maine, New Hampshire, Vermont, Public Works Administration. Letter to Lucius D. Barrows, Chief Engineer, Highway Commission, at Augusta, Maine. July 16, 1935.

Maine-New Hampshire Interstate Bridge Authority. Clipping file, located in Bridge Maintenance Office, Kearsarge Way, Portsmouth, N.H.
—“Draft of a Proposed Plan of Procedure for Use After 11/1/72.”
—Minutes of Regular and Special Meetings (I.B.A. Minutes). 1937-present.
—Seventeenth Annual Report. March 15, 1973.
—“Specifications for Painting...” Filed with Maine Department of Transportation.

Morrow, Raymond E., Executive Secretary of the Bridge Authority. Letter to A.R. Sirois, Bridge Maintenance Engineer, Maine Department of Transportation. January 18, 1971.
—Letter to Maynard L. Young, Painting Contractor. July 9, 1968.

New Hampshire, State of. *Eighteenth Annual Report of the State Highway Department of New Hampshire.* For the Period February 1, 1940 to January 31, 1941, inclusive. Concord, N.H.: N.H. State Highway Dept., 1942. Pp. 17-19: The Maine-New Hampshire Interstate Bridge Authority, with a photograph of the bridge. Courtesy of John Moore, N.H. Dept. of Transportation, Bridge Design Section.
—Extracts from Public Laws Relating to Highways, Bridges, Etc. Revised 1938. Concord, N.H.: N.H. State Highway Department, 1938.

Peverly, Elaine M. *Bridge Over the Piscataqua.* Sanford, Maine: Williams' Printers. Concerning the Piscataqua River ("High Level") Bridge, opened in 1972. Many construction photographs of this cantilever bridge. Four workers were killed in the collapse of a construction platform (monument below Maine approach) and one in a grading accident. The story regarding a construction death in the Memorial Bridge (opened 1923) appears to be apocryphal, and the author has been unable to confirm the report that a sandhog (pneumatic caisson worker) died of the bends in construction of the Interstate Bridge.

Plowden, David. *Bridges: The Spans of North America.* N.Y.: The Viking Press, 1974.

Rowe, John Frink. *Newington, New Hampshire.* Phoenix Publishing: Canaan, N.H., 1987. On p. 124 is reproduced a contemporary sketch of the (no longer extant) 1794 pile bridge in Newington, while p. 187 has a photograph of the 1873 pile railroad bridge from Newington to Dover, N.H. The latter was replaced in 1933 by the General Sullivan Highway Bridge, later part of the Spaulding Turnpike (1956), the southerly end of which is the Portsmouth Traffic Circle on the U.S. Route 1 Bypass. The Portsmouth-Dover rail line was abandoned.

Sverdrup & Parcel and Associates, Inc. *New Hampshire Historic Bridge Inventory: Portsmouth Bridge No. 247/084.* Boston: September 1982. Summary information and statistics on the Memorial Bridge, Portsmouth, N.H., to Kittery, Maine.

Thompson, Mary P. *Landmarks of Ancient Dover*. Durham, N.H.: Durham Historic Association, 1965 (reprint of the 1892 edition). Pp. 193ff.: "PASCATAQUA BRIDGE, otherwise PISCATAQUA," regarding the 1794 pile bridge from Fox Point in Newington, N.H., to Meader's Neck (Cedar Point) in Durham, N.H. Similar in important respects (length, depth, draw span) to the predecessor of the Interstate Bridge.

Tripp & Co., Inc. Prospectus for bond issue of $2,100,000: The Maine-New Hampshire Interstate Bridge Authority, 2.10% Bridge Revenue Refunding and Improvement Bonds. February 1, 1955.

Ungar, Dr. Eric E. Letter concerning vibration problem to David V. Dickey, N.H. Department of Public Works and Highways. August 27, 1976.

United States Army Corps of Engineers, New England Division. "Portsmouth Harbor, Piscataqua River Navigation Improvement Project; Preliminary Design Studies; Alteration of Maine-New Hampshire Interstate Bridge." Cleverdon, Varney & Pike, Inc. (Boston, Ma.) and Harrington & Cortelyou, Inc. (Kansas City, Mo.), Consulting Engineers; March 15, 1984.

United States Department of Commerce. (with state highway departments). *Factors Affecting Navigation Clearances for Highway Bridges in New England, New York, and New Jersey.* Washington, D.C.: U.S. Government Printing Office, 1955.
—*Navigational Clearance Requirements for Highway and Railroad Bridges.* Washington, D.C.: U.S. Government Printing Office, February 1955.
—*United States Coast Pilot, Atlantic Coast Section A.* 5th.ed. Washington, D.C.: U.S. Government Printing Office, 1950.

Waddell, J.A.L. *Bridge Engineering*. 2 vols. N.Y.: John Wiley & Sons, 1925.

White, Tom C., and A.R. Sirois, Bridge Engineers, Maine Highway Department. Letter to Raymond E. Morrow, Executive Secretary of the Bridge Authority. March 26, 1968.

Whitehouse, Robert A., and Catherine C. Baudoin. *The Port of Dover: Two Centuries of Shipping on the Cocheco.* Portsmouth Marine Society. Portsmouth, N.H.: Peter E. Randall, 1988. For the controversy surrounding the building of the 1822 Portsmouth bridge, see Chapter 3: "The Town of Dover v. the Portsmouth Bridge."

Wilder, Max L., Bridge Engineer, Maine Highway Department. Letter to F.E. Everett, New Hampshire Commissioner of Highways. February 25, 1937.

—Letter to Aldo A. Bartlett, Junior Engineer Inspector, Public Works Administration, at Stonington, Maine. December 30, 1938.

Interviews.

Abel, James. Retired bridge operator. Kittery, Maine. October 9, 1987.

Allard, Robert A. Assistant Treasurer of the Authority. Allard's North American Moving and Storage, Newington, N.H. October 14, 1987.

Barnard, Everett B., Bridge Engineer, Maine Department of Transportation. September 2, 1987.

Clark, Norman. Portsmouth, N.H. November 25, 1987. Portsmouth bypass area.

Cobb, Chandler B. Kittery, Maine. October 20, 1987. Railroads.

Conner, Dr. Ernest, Director, New Hampshire State Port Authority. September 1987.

Conners, Dana F., Maine Commissioner of Transportation and Chairman of the Bridge Authority. September 2, 1987.

Fitzsimons, Gray. HABS/HAER Division of U.S. National Park Service, Washington, D.C. September 30, 1987. Bridge bibliography.

Foley, Clarence, Bridge Operator. October 27, 1987.

Grinstead, Willis G., Harrington and Cortelyou, Consulting Engineers. July 20 and September 8, 1987.

Hancock, Frank, Attorney and Counsel for the Bridge Authority, Strater, Hancock and Erwin; York, Maine. August 27, 1987.

Hanscom, Samuel ("Buzzy"). Hanscom's Truck Stop, Portsmouth, N. H. November 16, 1987.

Holt, Captain Norman Y. Kittery, Maine. November 5, 1987.

Holt, Captain Richard. Portsmouth Navigation Co., Portsmouth, N. H. October 6 and October 23, 1987.

Howard, James, Bridge Operator. August 20, 1987.

Laderbush, Harry L., brother of George R. Laderbush. Portsmouth, N.H. November 4, 1987.

Long, Sarah Mildred. Portsmouth, N.H. July 29 and October 27, 1987.

Marconi, Michael. December 9, 1987. Re: construction rescue boat.

Newman, Edwin M., retired partner, Harrington and Cortelyou, Assistant Resident Engineer on the bridge. Corpus Christi, Texas. October 1, 1987.
—Letter to author. October 25, 1987.

Palfrey, Kennard. Portsmouth, N.H. November 27, 1987. Portsmouth bypass area.

Peoples, H. Denton. Taped recollections. San Marino, California. November 17, 1988 and January 8, 1988.
—Telephone interview. November 14, 1987.

Perry, Rita A., member of the Authority. Kittery Point, Maine. October 15, 1987.

Riley, Leland, Jr. Kittery, Maine. October 29, 1987. Regarding Kittery Junction and Kittery bypass.

Searles, Ernest, Maintenance Engineer for the New Hampshire Department of Public Works and Highways. August 26 and December 4, 1987.

Silverman, Harold J., Clerk of the Authority. Portsmouth, N. H. August 11, 1987 (his friends call it "Harold's Bridge").

Staples, Roger W., Treasurer of the Authority. Eliot, Maine. October 12, 13, and 29, 1987.

Starkey, Charles, Bridge Operator. October 15, 1987.

Stevens, David H., former member of the Authority, retired Maine Commissioner of Transportation and recently retired Chairman of the Maine Turnpike Authority. Hallowell, Maine. August 15, 1987.

Stickney, Wallace F., New Hampshire Commissioner of Transportation and Vice-Chairman of the Authority. August 25, 1987.

Walsh, Paul, Walsh Brothers Movers, Kittery, Maine. November 6, 1987. Kittery bypass, traffic circle.

Whitaker, Robert H., former member of the Authority and New Hampshire Commissioner of Transportation. Concord, N.H. August 25, 1987.

Williams, Robert, retired Bridge Operator. Kittery Point, Maine. November 27, 1987.

Young, Hiram A., Supervisor of Maintenance for the Interstate Bridge. August 13 and October 8, 1987.

Index

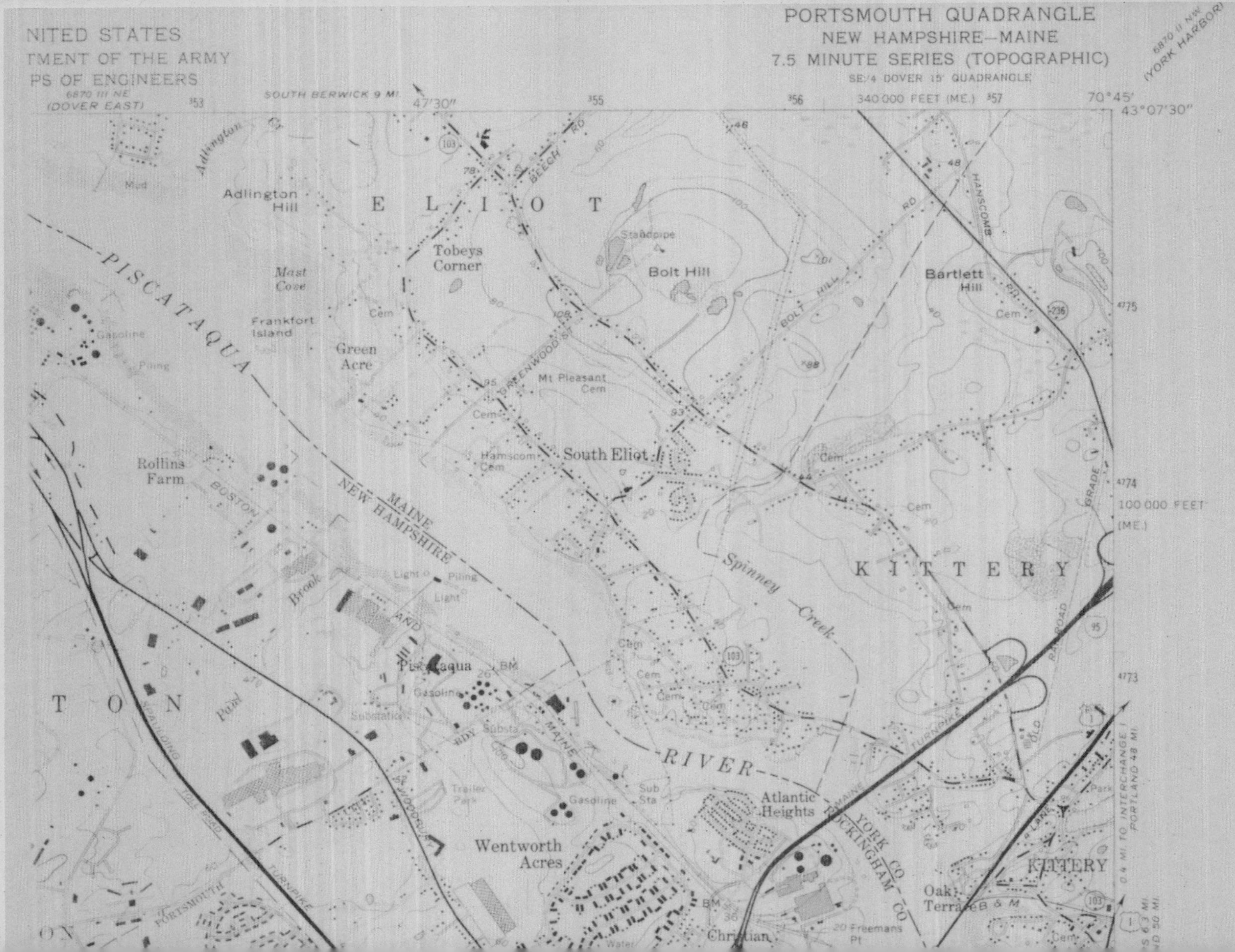
NITED STATES
TMENT OF THE ARMY
PS OF ENGINEERS
6870 III NE
(DOVER EAST)
PORTSMOUTH QUADRANGLE
NEW HAMPSHIRE—MAINE
7.5 MINUTE SERIES (TOPOGRAPHIC)
SE/4 DOVER 15' QUADRANGLE
6870 II NW
(YORK HARBOR)
SOUTH BERWICK 9 MI.
47'30"
340 000 FEET (ME.)
70°45'
43°07'30"
100 000 FEET
(ME.)
E L I O T
K I T T E R Y
T O N
PISCATAQUA
RIVER
MAINE
NEW HAMPSHIRE
YORK CO
ROCKINGHAM CO
Adlington Hill
Tobeys Corner
Mast Cove
Frankfort Island
Green Acre
Bolt Hill
Bartlett Hill
South Eliot
Mt Pleasant Cem
Hamscom Cem
Rollins Farm
Spinney Creek
Piscataqua
Atlantic Heights
Wentworth Acres
KITTERY
Oak Terrace
Christian
Freemans Pt
HANSCOMB
BEECH RD
GREENWOOD ST
BOLT HILL RD
BOSTON AND MAINE
SPAULDING TOLL ROAD
TURNPIKE
RAILROAD GRADE
Standpipe
Gasoline
Substation
Trailer Park
0.4 MI. TO INTERCHANGE 1
PORTLAND 48 MI.